Evolution Since Coding

Evolution Since Coding

Cradles, Halos, Barrels, and Wings

Zachary F. Burton
Michigan State University
East Lansing, MI, United States

Academic Press is an imprint of Elsevier
125 London Wall, London EC2Y 5AS, United Kingdom
525 B Street, Suite 1800, San Diego, CA 92101-4495, United States
50 Hampshire Street, 5th Floor, Cambridge, MA 02139, United States
The Boulevard, Langford Lane, Kidlington, Oxford OX5 1GB, United Kingdom

Notices
Knowledge and best practice in this field are constantly changing. As new research and experience
broaden our understanding, changes in research methods, professional practices, or medical treatment
may become necessary.

Practitioners and researchers must always rely on their own experience and knowledge in
evaluating and using any information, methods, compounds, or experiments described herein. In
using such information or methods they should be mindful of their own safety and the safety of
others, including parties for whom they have a professional responsibility.

To the fullest extent of the law, neither the Publisher nor the authors, contributors, or editors,
assume any liability for any injury and/or damage to persons or property as a matter of products
liability, negligence or otherwise, or from any use or operation of any methods, products,
instructions, or ideas contained in the material herein.

Library of Congress Cataloging-in-Publication Data
A catalog record for this book is available from the Library of Congress

British Library Cataloguing-in-Publication Data
A catalogue record for this book is available from the British Library

ISBN: 978-0-12-813033-9

For information on all Academic Press publications visit our website at
https://www.elsevier.com/books-and-journals

Working together
to grow libraries in
developing countries

www.elsevier.com • www.bookaid.org

Publisher: Mica Haley
Acquisition Editor: Kristi Gomez
Editorial Project Manager: Pat Gonzalez
Production Project Manager: Kirsty Halterman and Karen East
Designer: Alan Studholme

Typeset by TNQ Books and Journals

Contents

Preface   ix
Introduction   xi

1.   Big Questions   1
2.   What This Book Is About   3
3.   Scientific Acronyms   7
4.   Molecular Graphics   9

Section I
The Old Testament of Gene Regulation

5.   Evolution as a Cutout Doll Problem I   13
6.   Evolution as a Cutout Doll Problem II   17
7.   The Best Art   21
8.   Intelligent Design   23
9.   Polymers   25
10.   LEGO (Trademark) Life   27
11.   α/β Proteins   33
12.   The Inevitability of α/β Folds   45
13.   Evolution Puts the Y (Why) in Biology   49
14.   Evolution Is Not Anti-Religion   51
15.   Computers in Biology   53
16.   The Old and New Testaments of Gene Regulation   55
17.   Evolution as a Cutout Doll Problem III   59
18.   Multi-Subunit RNA Polymerases Book I   65
    Double-Ψ–β-Barrels (The Active Site)   67
    The Bridge Helix   68

	The Trigger Loop	68
	σ Factor Binding to Promoter DNA	68
	Zn1 and Zn2	69
	Do Not Try This at Home	71
19.	Multi-Subunit RNA Polymerases Book II	73
20.	The Chemical Synthesis of Life	83
21.	The RNA World	85
22.	Ribosomes	87
23.	The RNA-Protein World	89
24.	Transfer RNA	91
	Generation of Sequence Repeats	103
	"Intellectual Property" in Biology and Evolution	104
	Try This at Home	104
	Now Try This	105
25.	The Three Domains of Life on Earth and Scientific Working Hypotheses	107
	Last Universal Common (Cellular) Ancestor	107
	Archaea	109
	Bacteria	109
	Eukaryotes	110
	Descriptions and Predictions	111
26.	LUCA	113
27.	General Transcription Factors and Promoters	115
	Try This at Home	121
28.	Archaea	123
29.	Bacteria	125
30.	Methane and Oxygen	129

Section II
The New Testament of Gene Regulation

31.	LECA	133
32.	Eukaryotic Multi-Subunit RNA Polymerases, General Transcription Factors, and the CTD	139
33.	Patchwork Eukaryotic Phylogenomics	145
34.	Plants	147
35.	The Permian–Triassic Extinction	149

36. The Triassic–Jurassic Extinction   151
37. The Cretaceous–Paleogene Extinction   153
38. The Paleocene–Eocene Thermal Maximum   155
39. Promoter Proximal Pausing and the CTD Interactome   157
40. Human Evolution   161
    Of Mice and Men   162
41. Human Cancer   165
42. Homology Modeling and Cryoelectron Microscopy   169
43. Human Extinction   171
44. Evolution Versus Faith   173
45. Concept in Biology   175
46. Other Books and Studies   177
    References   181

Supplementary Materials   185
Index   187

Preface

Public discussions of evolution versus "intelligent design" or a deity are no more enlightening than public discussions of man-made global warming and ocean acidification with an oil executive. Basically, one is having the wrong conversation with the wrong people. Evolution is no more anti-religion than wristwatches or jet airplanes. As recognized by Arrhenius (late 1800s to early 1900s), carbon dioxide is a greenhouse gas that dissolves in water to produce acid. So, why argue about that? What evolution appears to do is to render a deity unnecessary, at least, for the genesis of life on earth, a story that now can be told with surprising details, confidence, and authority.

I wrote this book to tell a simple, brutal story of genesis that integrates RNA structure/function, protein structure/function, RNA synthesis, protein synthesis, metabolism, and phylogenomics. As living organisms, our story is rich and wonderful, and I hope I tell some of our most ancient stories well. I have certainly learned a lot writing this book. Perhaps regrettably, my best studies in the scientific literature were written based on my ancient evolution studies of transcription systems and cloverleaf tRNA. This book is an attempt to reach a larger, perhaps less-specialized but more enthusiastic audience. I have great passion for this book and subject.

So, why might you be interested in *Evolution Since Coding*?

Although you may not think so, you wish to learn molecular graphics, and I will encourage you to do so. Initially, low-level proficiency with molecular graphics programs takes a few hours. Of course, ultimately, this core life skill will cause you much frustration, anguish, and pain. But, if you do not want a life of thwarted dreams, you should avoid biology.

I tell a very straightforward story of evolution of life on earth. I am amazed at how simple, integrated, and predictive the story of genesis has become.

This is a story of code breaking. The codes are dumb, but not so dumb that scientists could not break them. Genetic codes provide a written and historic record of evolution, and much of the initial print remains legible.

Transcription (RNA synthesis) systems resemble a Rube Goldberg construction with seemingly endless add-ons and compensations. Evolution makes surprising sense out of otherwise incomprehensible transcription systems.

Translation (protein synthesis) systems comprise a Rube Goldberg contraption with a distinct history. Evolution of transfer RNA (tRNA) provides the core narrative and makes conceptual sense out of translation systems.

The evolution of metabolism is conceptualized. A simplified evolutionary description may be a useful complement to the perplexing treatment presented in biochemistry courses. When you realize that much of metabolism is based on two protein folds of the same basic pattern ($(\beta-\alpha)_8$, TIM barrels, and Rossmann folds), the subject may seem conceptually less daunting.

The evolution of three domains of life on earth (archaea, bacteria and eukaryotes) reduces to an instructive working model. To a remarkable extent, evolution of the three domains tracks evolution and divergence of transcription systems: that is multisubunit RNA polymerases and their general transcription factors.

Eukaryote complexity is the consequence of archaeal and bacterial fusions. The clash of many prokaryotic genomes to form eukaryotes generates organismal complexity in understandable ways.

So, if you care about genesis of life on earth, this book is for you.

If you care about the RNA-protein world, this book is for you.

If you care about transcription and translation systems, this book is for you.

If you care about metabolism, this book is for you.

If you care about eukaryotic complexity, this book explains that.

If you care about cancer, viral infection, and/or animal complexity, I offer simple foundational models.

Despite seemingly endless possibilities (and current politics), evolution is frighteningly conservative. Significantly, evolution appears to lack a strong ethic of innovation and instead builds new functions on existing scaffolds. Very often, generation of biological complexity involves iteration of motifs followed by folding and then coevolution of partners. Repeatedly, evolution appears to be a race to fold, to closure, and to resulting solubility with later emergence of refined biological function and specialization. Intricate folds and much larger biological systems are duplicated and repurposed.

So, this is a conceptual story of genesis supported by many beautiful color pictures and verifiable hypotheses. Molecular graphics is high art, and soon you will make your own. I hope no one is too offended or too upset by the simplicity. Personally, I find the models in my book a little frightening and unsettling.

Introduction

Evolution Since Coding is an updated (as of 2017) version of the story of genesis (since coding), written from a scientific point of view. I did not attempt a full description of the chemical synthesis of life, because I cannot now tell this fascinating and important story with any semblance of authority. Currently, I am not certain that anyone can tell this story and be convincing. I also largely steer clear of the central issues of cell energetics and the encapsulation of organisms and energy-generating systems within cells. Rather, I wanted to tell the short and brutal story of evolution of living systems based on a reading of the genetic code and protein and RNA structure/function. Significantly, the genetic code is a preserved written history of the evolution of life on earth. No deity is invoked because no deity appears to be required.

I wrote this book because I spent my life working on transcription systems, which, from a human point of view, seem bewilderingly complex and poorly designed. With their endless add-ons and compensations, RNA synthesis systems appear to be relics of evolutionarily design. Overly ornate RNA synthesis systems therefore make much better sense when viewed through an evolutionary lens, which is built from an emerging integration of phylogenetics and protein and RNA structure/function. Considering these aspects, RNA polymerases, general transcription factors, gene-specific transcription factors, and promoters reduce to straightforward working models. No one was as surprised as I was that transcription systems could allow for such simple conceptual explanations.

When I thought I was beginning to understand transcription, evolution of the three domains of life took shape for me. Remarkably, life on earth breaks into a simple diagram of the RNA-protein world going to a DNA genome world, diverging from LUCA to archaea and bacteria, and then fusing of archaea and bacteria to generate eukaryotes. Analysis of transcription systems provided huge insight into these transitions, and a confluence of exploding phylogenomics and transcription studies appears to largely describe the evolution of living systems.

I had never known much about protein synthesis, but I felt that I had to try to learn. Translation was the missing link in my story, because if you understand transcription and translation, you appreciate the RNA-protein world, which is the bedrock of genesis. I was certain that DNA replication came later, starting with a retroviral mechanism (transcription followed by reverse transcription) at LUCA (the last universal common (cellular) ancestor). Separate DNA replication systems evolved independently in archaea and bacteria after divergence.

I hoped that insights from evolution of transcription systems might carry over to translation systems, and the evolution of transcription systems is starkly simple. Transcription systems are built from a small number of LEGO pieces often by iteration and coevolution. In the most ancient evolution of life on earth, repetition of motifs appears again and again, as if this is a (or the) primary mechanism to evolve enduring biological complexity. Repeats leave a print that remains discernable, and detection of ~4 billion-year-old sequence repeats provides a means to code-break evolution. Remarkably, evolution of metabolic systems can also be described by a mechanism of iteration and coevolution.

Although I still do not know as much about translation as transcription, conceptually, the problems are very similar. Transfer RNA (tRNA) appeared to be the central function in protein synthesis around which the rest of the overly ornate translation system evolved. Therefore, if evolution of tRNA could be understood, the problem of evolution of protein synthesis could be addressed. Cloverleaf tRNA is an iconic molecule and one of the first lessons in biochemistry courses. I had little hope that I could solve the tRNA puzzle, but, with my colleagues, I tried anyway. As it happens, tRNA can be solved like a crossword puzzle. I am lousy at crossword puzzles, and they give me a headache. Sadly, it took me a few months to solve cloverleaf tRNA evolution, so this effort caused many headaches. With a little guidance, you will solve this puzzle much more quickly, and you will wonder what took me so long. Understanding tRNA evolution provides a conceptual working model for evolution of translation systems. Similar to transcription systems, tRNA is a simple story of iteration, folding, and, in this case, internal, symmetrical RNA processing.

If you understand transcription and translation, you begin to understand the RNA-protein world. If you understand retroviral replication, you progress to LUCA and the first cells with DNA genomes. If you understand RNA polymerases, general transcription factors and promoters, you understand divergence of archaea and bacteria. If you then figure out evolution of eukaryotes, you know a lot of biology. So, how did eukaryotes evolve? Strangely, eukaryotes are a complex genetic fusion of multiple archaea and multiple bacteria, and this story, although simple, is too good to be spoilt in the *Introduction*. Eukaryotic complexity arises from this fusion and from many modifications to transcription systems that are coincident with the birth of eukaryotes. Phylogenomics studies based on massive DNA sequencing projects have contributed huge insights into eukaryote evolution, identifying and connecting many formerly missing links. Eukaryogenesis is a very good story that can now be told with confidence. Many important details will continue to emerge to enhance this story, but it appears that the working model is reasonably complete. Eukaryotic complexity (compared to prokaryotes) can be described in terms of the tortured path of serial prokaryote within prokaryote fusions, so models are strongly predictive. Models for eukaryote complexity also help to describe cancers and viral infections.

To fully appreciate this book, you will have to teach yourself molecular graphics. If you do not do this, you will not appreciate iteration and coevolution.

You will not appreciate protein and nucleic acid structure and function. Everyone interested in biology must learn molecular graphics, not just you. Think of this as a fun computer game that you will never win, so you are doomed to daily frustration with little reward or payout. You can happily start reading the book without knowing any molecular graphics. With graphics, however, you can confirm and visualize the hypotheses I put forward in the book.

So, this is a story whose time has come. From my point of view, genesis is a big story with far-reaching implications. Evolution, of course, puts the "y" (why) in biology. This is a deeply conceptual story of the evolution of life on earth, and most working models presented in *Evolution Since Coding* will stand scrutiny and should stand the test of time. To dig deeper, read the scientific literature and maybe even one of my papers. I would be honored if you did.

Chapter 1

Big Questions

So, what are the big questions about how life evolved on earth (or somewhere in space)?

What are the processes for chemical evolution of pre-life before coding? How do you get to an RNA world? How do you couple energy transduction (redox or chemical energy) to polymerization?

Molecular biologists are good with coding. Once you get to coding, the story makes much better sense.

How can one imagine the RNA world? This is a strange world that is difficult to grasp and visualize. Consider LEGO (trademark) life, basket weaving, cutout dolls and protein origami. There may never have been an RNA world without protein. There may have been an RNA-protein world in which RNA was more dominant.

How do you get from the RNA world to RNA-protein world? How is the ribosome invented?

How might one consider the ribosome?

Why did the ribosome stay a ribozyme (an RNA enzyme) for peptidyl transfer? Why did a protein not assume ribosome ribozyme function?

Are there patterns in modern day proteins that give hints to ancient evolution? Yes! I say yes!

Once you get to the RNA-protein world, the world begins to look like life. Evolution looks like evolution. Proteins are synthesized on ribosomes or proto-ribosomes. Before long, as catalysts, proteins begin to displace most of the ribozymes. RNA genomes appear strange and fragmented but somewhat recognizable. Many metabolism and energy transduction systems evolved in the RNA-protein world, possibly prior to last universal (cellular) common ancestor (LUCA).

How do you get from the RNA-protein world to the DNA-genome world and what are the consequences? How do you get to the LUCA: the last universal common (cellular) ancestor of bacteria, archaea and eukaryotes?

How did bacteria and archaea diverge from LUCA? This is a big question.

How did bacteria and archaea fuse to generate eukaryotes? What might be the mechanism to evolve LECA: the last eukaryotic common ancestor?

How important is endosymbiosis in horizontal gene transfer and genesis of eukaryotic cellular organelles?

Where did the mitochondria come from? You may know the answer to this question.

Evolution since Coding. http://dx.doi.org/10.1016/B978-0-12-813033-9.00001-9

Where did the cell nucleus come from? What drove evolution of the cell nucleus?

Where did the chloroplast come from? How did plants evolve?

How important is horizontal gene transfer and what are the mechanisms?

How can one imagine endosymbiosis?

How important are viruses in horizontal gene transfer?

How can selfish DNA and RNA elements act as agents for horizontal gene transfer?

How does the virus/selfish DNA/RNA element world interface with the cellular organism world?

How does the biosphere affect the atmosphere?

How does the geosphere affect the atmosphere?

Why are eukaryotes complex compared to prokaryotes?

Why and in what ways are eukaryotes different from bacteria and archaea?

Why do eukaryotes have complex genomes? Why do eukaryotic genomes generally appear less streamlined and efficient than bacterial and archaeal genomes?

What are the gene regulatory mechanisms that license eukaryote complexity? How are gene regulatory networks specialized and evolved?

How did nerves evolve?

What about skeletons?

How did vertebrates evolve?

How did humans evolve?

What is distinct about humans?

What effects do humans have on the biosphere?

How do humans interface with the virus/selfish element world?

Why do complex eukaryotes suffer cancer?

How does a cancer evolve? What does cancer teach us about evolution?

Are organisms evolved or intelligently designed?

What limits the complexity of living organisms? Why are organisms complex? How complex can organisms become?

If you wish to think like a scientist, think in terms of questions. By comparison, answers are boring.

Chapter 2

What This Book Is About

The intention of this book is to provide a simple, verifiable understanding of evolution of life on earth. The argument, at its core, is frighteningly simple: (1) our "designs" are surprisingly simple and not obviously "intelligent"; (2) RNA synthesis describes evolution; (3) evolution describes RNA synthesis; and (4) evolution of seemingly overly complex protein synthesis systems can be simply modeled. Therefore, evolution can be viewed by understanding mechanisms of RNA and protein syntheses. Before DNA genomes, life passed through a strange RNA-protein world. RNA-template-dependent RNA polymerases of the two double-Ψ–β-barrel types (described in detail in this book) became the dominant enzymes for RNA synthesis. As life evolved into the DNA-template world, these enzymes evolved to become DNA-template-dependent RNA polymerases and spread to all cellular life. RNA, therefore, was and is the central coding molecule of life on earth and remains the intermediary between DNA genomes and protein catalysts. Because life comes from an RNA-protein world, mechanisms for RNA synthesis remain at the heart of life processes, and understanding RNA synthesis is a key to understand evolution.

I am an expert in RNA synthesis mechanisms in bacterial and eukaryotic systems. After listening to a talk by Seth Darst (Rockefeller University) on the evolution of multi-subunit RNA polymerases (2013), I became aware of the importance of evolutionary analyses to understand the mechanisms of RNA synthesis. Darst encouraged me to review the work of Aravind (NCBI, National Center for Biotechnology Information) and Eugene Koonin (NCBI) on the evolution of RNA polymerases, which led me to read more broadly in the vast realm of core protein motifs and phylogenomics. My book is intended to bridge the fields of RNA synthesis, phylogenomics, ancient evolution, and core protein motifs. The book is based on the theme that RNA synthesis describes ancient evolution and ancient evolution describes RNA synthesis. Furthermore, core protein motifs are preserved since the RNA-protein world (~4 billion years ago), so a language of the initial evolution of life on earth can be read in genetic code. These unifying insights bring startling concept to biochemistry, RNA synthesis, evolution, and metabolism.

Furthermore, to paraphrase J.R.R. Tolkien, this is a book that grew in the telling. To grasp ancient evolution primarily requires understanding two processes: RNA synthesis and RNA-encoded protein synthesis. DNA, of course, is an evolutionary afterthought, an improvement on RNA as a repository for genetic information. So, working on this book, I decided to see whether I could begin to understand protein synthesis at least at the working model level. I was fairly satisfied with my understanding of the evolution of RNA synthesis mechanisms.

Of course, one would like to also understand how cell membranes arise, evolution of energy transduction systems, and the genesis of signaling systems. Some of these very complex stories are partially embedded in an understanding of transcription (RNA synthesis) and translation (protein synthesis) systems. Membrane, bioenergetics, and signaling systems are discussed but are not the central focus of this book.

To understand protein synthesis requires focusing on transfer RNA (tRNA). As an evolved entity, the ribosome appears to be incomprehensibly complex, so a simple working model for evolution of translation systems is initially difficult to imagine. This problem is largely solved, at least in principle, through recognition that tRNA is the core function, and seemingly overly ornate translation systems can largely be built up and evolved around tRNA. With my colleague, Robert Root-Bernstein (MSU; Physiology), I addressed this issue by solving the evolution of tRNA. This story is surprisingly similar to the evolutionary histories of transcription systems. Ancient evolution follows simple governing principles: largely iteration of simple motifs followed by coevolution of partners. In many cases, the iterated pattern remains and can be read.

This book is not meant to be a comprehensive textbook on evolution. Rather, an attempt is made to make a focused and comprehensible argument that anyone can verify. Many important discoveries about evolution, therefore, are not included or are treated briefly. My effort was to provide sufficient detail on ancient core protein motifs and RNA synthesis mechanisms as they relate to telling a believable and compact story of the genesis and evolution of life on earth. The story about tRNA is a late add-on to the text that turns out to fit the central narrative. At some level, this is a *how to evolve life on earth* book, in which the reader is encouraged to participate.

The problem of "intelligent" versus evolutionary "design" is addressed from analysis of ancient core protein motifs. Remarkably, there are core protein motifs that are ~4 billion years old on an ~4.6 billion year old planet. A human, therefore, with a limited lifespan can see back ~4 billion years by looking at "immortal" core protein motifs. Because we are constructed of LEGO (Trademark) pieces or cutout paper dolls, our construction is not "intelligent," and there is no necessity to invoke the existence of a deity to explain life on earth. Remarkably, life appears to be evolved rather than to be intelligently designed. Life appears to reduce to a problem of linking energy transduction to the formation of polymers that eventually become coding and self-replicating polymers.

From a primordial soup of core protein motifs, RNA synthesis was born. Life comes from a strange RNA-protein world, and the central stories of life on earth, therefore, are the stories of RNA synthesis and templated protein synthesis. Although the story of genesis of RNA synthesis systems is chronicled in the scientific literature, this is a story that has not been disseminated to a broad popular audience. The story of protein synthesis follows easily from the story of RNA synthesis. The story of genesis is important to tell, because the story forms the core of the human soul. If not the story of genesis, what might be of enduring interest or importance? What, after all, is the meaning of life on earth?

In any event, this is my short, brutal tale of evolution of life on earth. I try to tell much of the story with pictures generated using molecular graphics, so the reader is encouraged to check my work using molecular graphics. Anyone can learn these tools.

Chapter 3

Scientific Acronyms

If, for any reason, the reader is disappointed by a paucity of incomprehensible scientific jargon or unfamiliar acronyms in this book, the author sincerely apologizes. Where necessary, I shall try to provide subtitles to science speak. If the reader finds words or phrases that seem incomprehensible, Wikipedia can provide a description. Otherwise, please be assured that the author knows many arcane scientific acronyms and strange words that I shall try to refrain from using. That being said, there is significant complexity to RNA synthesis systems and their networks, making some jargon unavoidable. Please, however, look beyond the language and relate to proteins from their structures and core motifs. Proteins are Frankenstein's monsters, and parts are parts. There are many and striking protein family resemblances that, using molecular graphics, anyone can see.

In some cases, because of the long passage of time, protein similarities can no longer be identified by reading linear amino acid sequence. In many of these cases, however, similarities are still apparent from protein secondary and tertiary structures, some of which are preserved over ~4 billion years of evolution on earth. Therefore, some understanding of protein structure will have to be attained by the reader to understand this book, and an attempt will be made to help with this.

To simplify, what all this means is that, using conserved protein secondary and tertiary structures, much of evolution of life on earth becomes a simple problem in iterated pattern recognition that anyone can discern. No PhDs, or MSs, or BSs, or special knowledge of any type are initially required. Anyone can do this. This is a simple problem in code breaking. The code is simple, and the code has been broken. A purpose of this book is to demonstrate the breaking of the code.

Think of DNA, RNA, and protein sequences as a written language. Think of the ribosome as a Rosetta Stone for translating messenger RNA sequences. Think of protein secondary and tertiary structures as part of the language. According to this view, there is a readable history of life on earth written in genetic code, much of which was written ~4 billion years ago. I shall show you how this language can be read.

Evolution since Coding. http://dx.doi.org/10.1016/B978-0-12-813033-9.00003-2

Chapter 4

Molecular Graphics

The story of genesis cannot be appreciated unless the reader utilizes molecular graphics.

Molecular graphics programs are freely available and easy and rewarding to use.

I could teach you to use molecular graphics programs in about an hour. Using tutorials, you can teach yourself to use these programs in about the same amount of time.

Unless you learn molecular graphics, you will not be able to confirm the information content of this book. By contrast, if you learn molecular graphics, you can test my hypotheses, you will learn protein chemistry and you will learn biology and evolution.

If you fail to learn molecular graphics, please do not criticize my book.

If you learn molecular graphics, and you still do not believe me, feel free to criticize as much as you like, and please be vocal and public in your criticism. All publicity is good.

Evolution since Coding. http://dx.doi.org/10.1016/B978-0-12-813033-9.00004-4

9

Section I

The Old Testament of Gene Regulation

Chapter 5

Evolution as a Cutout Doll Problem I

Protein structure/function may seem to be an arcane branch of human knowledge, but I assure you it is not.

Protein structure and function can be understood and analyzed by anyone.

The evolution of protein structures and functions can also be understood in significant detail.

A working model for the genesis of life on earth is available, due to the efforts of many. I would name the following, from whom I particularly take inspiration: (1) Eugene Koonin (NCBI); (2) Aravind (NCBI); (3) Andrei Lupas (Max Planck Institute); (4) Seth Darst (Rockefeller University); (5) Carl Woese (University of Illinois at Urbana-Champagne); (6) Lynn Margulis (University of Massachusetts at Amherst); (7) James Lake (UCLA); (8) Finn Werner (University College London); (9) many others.

To a surprisingly great extent, ancient evolution can be visualized back to the inception of cellular and DNA-based life and beyond, to the RNA world and the RNA-protein world.

The reason that ancient evolution can be visualized is that core protein motifs that existed at the inception of life still exist and can be traced back near to their origins. Therefore, the inception of life can be visualized in extant core protein motifs. Just to be clear, the inception of DNA genomes and cells is thought to have occurred more than 3.5 billion years ago, and the earth is only about 4.6 billion years old. This barely gives the planet time to cool before the inception of life, and, because of the early instability of our solar system, the first 1 1/2 billion years on earth were particularly rough and chaotic.

In this book, I refer to LUCA: the last universal cellular common ancestor. Think of LUCA as the first cellular organism and the first DNA genome organism. To me, LUCA is more a concept than an organism, and I will make no serious attempt at an accurate portrayal of LUCA and, rather, leave this important discussion to evolutionary biologists and experts in phylogenomics. I come to this topic late in life and from the perspective of multi-subunit RNA polymerases, general transcription factors, promoters, and core protein motifs. From my point of view, the essential aspect of LUCA is its form

Evolution since Coding. http://dx.doi.org/10.1016/B978-0-12-813033-9.00005-6

at the divergence of archaea and bacteria. At this ancient time, LUCA was certainly cellular and possessed a largely intact DNA genome. LUCA is a beautiful concept whether or not its precise constituents and makeup can be divined. To me, the most essential aspect is the great divergence to archaea and bacteria.

Before LUCA, ancient evolution becomes more opaque and more mysterious, but relics of earlier times remain, primarily in core protein motifs, ribosomes, transfer RNA, ribosomal RNA and other ribozymes (RNA enzymes). The RNA-protein world is a fascinating concept to consider, as is its evolution to LUCA. Think of the RNA-protein world as about 4 billion years ago on a ~4.6 billion-year-old earth.

Before the RNA-protein world, consider an RNA world in which ribozymes are more plentiful and initially more talented than many protein enzymes.

Before the RNA world, consider the chemical evolution of life, if you can and will. I find this topic mostly beyond my ken and will not treat it fully in this book. Generation of life after coding, however, provides insights into prebiotic times. Life is a story of soluble self-replicating polymers supported by energy transduction [redox (oxidation-reduction) and chemical energy]. These same rules must apply before the RNA world.

Moving forward from LUCA, consider the divergence of archaea and bacteria. Divergence is one of the major sign posts in evolution of life on earth. According to the dating of microfossils and stromatolites (residues from aggregated bacteria/archaea), divergence may have happened ~3.5 to 3.8 billion years ago. I shall argue that divergence can largely be explained from analysis of multi-subunit RNA polymerases, RNA polymerase general transcription factors, and RNA polymerase promoters (DNA sequences at which RNA polymerase binds to start transcription). Archaea and bacteria diverged because they developed distinct (but genetically related) strategies to interpret and replicate DNA genomes.

Based on evolutionary characterization, multi-subunit RNA polymerases are of two double-Ψ–β-barrel (double-psi-beta-barrel) types, a concept I shall explain in detail. In evolution, DNA-template-dependent multi-subunit RNA polymerases were preceded by RNA-template-dependent RNA polymerases of two double-Ψ–β-barrel types. Because life is derived from an RNA-protein world, RNA-template-dependent RNA polymerases precede DNA-template-dependent RNA polymerases, and evolution of RNA polymerases maps the course of life on earth since its inception, certainly since the RNA-protein world almost 4 billion years ago.

But humans are eukaryotes, and we have not gotten to eukaryotes yet. So, where do eukaryotes come from?

Eukaryotes appear to be an unholy fusion of archaea and bacteria, a kind of genetic patchwork. A major event was the endosymbiosis of an ancient

Lokiarchaeota archaea engulfing an α-proteobacterium resulting in fusion of genomes, catastrophic genetic violence, cell nuclei, and mitochondria. This cataclysmic clash of genomes unleashed almost unimaginable genetic aggression against the archaeal genome, perpetrated by a hopping bacterial selfish DNA element named the self-splicing "group II intron." Eukaryotes, therefore, are a remarkable and seemingly unlikely story of evolution and genetics.

And where do plants come from?

Similar to eukaryotes, plants also arose via endosymbiosis, in this case of a photosynthetic cyanobacterium occupying a primitive eukaryote. The cyanobacterium reproduced and ultimately left behind the chloroplast and relics of its genome within its host.

In addition to endosymbiosis, horizontal gene transfer is a potent mechanism for mobilizing genetic material between organisms. Horizontal transfer makes somewhat of a mess of an orderly tree of life by linking upper branches for the transferred genes. For many genes, the tree of life, therefore, may largely branch as a network rather than as a tree. Vectors for horizontal gene transfer include viruses and selfish DNA elements, some of which can transfer significant segments of DNA. It is worthwhile to consider that the currencies of evolution include not only separate organisms but also chunks of genes, individual genes, and core protein motifs.

And humans?

From the point of view of protein structure, core protein motifs, genes, promoters, genomes, evolution, etc., human evolution is not exceptionally interesting. Human evolution is a top-down problem more than a bottom-up problem. Here, I try to suggest the bottom-up approach of rebuilding evolution of life from parts. Life was generated from evolutionary assembly and selection of RNA-encoded core protein motifs. As it happens, humans have two double-Ψ–β-barrel-type multi-subunit RNA polymerases just like everybody else.

So, how can evolution be a cutout doll problem?

If you fold a piece of paper like an accordion and make scissors cuts to form mirror image dolls as the paper is unfolded, you have largely solved the problem of the evolution of core protein motifs (Fig. 5.1). You can make linear multimers of dolls, cradles of dolls, twisted multimers of dolls, or circles (halos or barrels) of dancing dolls. This is the level of complexity of ancient protein evolution. Ancient evolution reduces to a cutout doll problem that anyone can play.

FIGURE 5.1 Evolution is a cutout doll problem. Many ancient core protein motifs can largely be described according to the organization of β-sheets, which interact with neighboring β-sheets via hydrogen bonding (indicated by hand and foot holding of dolls). Numbers indicate the order of the β-sheet in the polypeptide chain. TIM barrels, Rossmann folds, and TOPRIM domains will be described in detail.

Chapter 6

Evolution as a Cutout Doll Problem II

The core concept of this book is that last universal (cellular) common ancestor (LUCA) and the RNA-protein world, ancient though they be, can be viewed in extant core protein motifs.

So, from Alaska, Sarah Palin can see Russia.

From your current vantage, you can see LUCA.

I shall show it to you.

In our quest for immortality, we are surrounded by and made up of the immortal.

By any human standard, core protein motifs are immortal. Core protein motifs go back to LUCA and beyond to the RNA-protein world. Astoundingly, ancient core protein motifs are ~4 billion years old on an ~4.6 billion year old earth.

So, why and how do core protein motifs resemble cutout doll patterns?

Core protein motifs are generated by simple genetic repetitions generated via duplications and/or gene fusions [i.e., RNA ligations (i.e., linking two identical RNAs together)].

Genetic duplications are very common genetic errors. Multiple repetitions of duplications result in multimerization of a motif. Multimerization (often of identical gene fragments) is a primary cause of genetic and protein structure complexity.

In the RNA and RNA-protein worlds, gene fusions, via ligation of two often identical RNAs, are hypothesized to be very common. This seems the most plausible mechanism to generate repeating sequences, which, in ancient evolution, appears to be a major driver of increased and enduring biological complexity.

The most prevalent and the most ancient protein motifs were generated via iteration of sequences. Although I shall show many examples, I shall not come close to saturating such a vast topic. But the fundamental concept remains important: biological complexity is very often generated through a simple and common genetic mechanism resulting in generation of iterated sequences that encode initially repetitive core protein motifs.

So, how are core protein motifs similar to cutout paper dolls?

To make this point, core protein motifs are generated as simple linear repeats folded into regular shapes: twisted sheets and barrels. The astounding

Evolution since Coding. http://dx.doi.org/10.1016/B978-0-12-813033-9.00006-8

Rossmann fold

TOPRIM domain

TIM Barrel

FIGURE 6.1   TIM barrels (8 up), Rossmann folds (8 up), and TOPRIM folds (4 up) modeled using cutout dolls. Hands and feet represent hydrogen bonds that link β-sheets.

observation remains that these simple core protein motifs have lasted forever (~4 billion years) without full degradation of the initial pattern.

Furthermore, considering rules of genetic coding and protein structure, these simple folds appear to be inevitable.

In one sense, ancient evolution was a race to form repeating core protein motifs—a race that the simplest, soluble, structured motifs were bound to win.

The race was on to form highly structured and soluble protein folds as quickly as possible. Ancient core protein motifs were the fastest stable and soluble structures to fold.

It is a race because more chaotic protein folds were not as fast to form the first structured and soluble enzymes and folds. For the most part, even today, new protein folds and motifs are rarely cut out of whole cloth. New protein structures, generally, are derived from existing folds with tested structural stability and good solubility.

I should pause here to admit that I do not fully understand why protein structures become immortal. For the moment, therefore, let us accept that this is true, because this certainly is true: core protein motifs can become immortal and can be preserved in evolution for billions of years. In subsequent chapters I shall try my best to rationalize this observation.

So, in the most ancient evolution, fastest is often best, as demonstrated through billions of years of unsuccessful competition against the most ancient structures.

To me, this observation is incredible. I had thought of evolution as being much more innovative. I could not have imagined that a core protein motif could

Double-Ψ–β-barrel

FIGURE 6.2 A cutout doll model representing core features of the double-Ψ–β-barrel (in this view, counterclockwise from back: (β1, β5) 2 up, (β2, β4) 2 down, (β3) 1 up, and (β6) 1 down).

become immortalized and last without replacement or substantial change or improvement through billions of years.

Ancient proteins were not cut out of whole cloth. Ancient proteins were generated by simple genetic iterations and folding. Think LEGO (trademark) life and protein origami. The idea of LEGO life is a chaotic primal soup of protein bits and RNA. Small protein LEGO pieces initially search for structure, solubility, and structural closure. What results includes cradle-loop barrels, RIFT barrels, double-Ψ–β-barrels, swapped-hairpin barrels, TIM barrels, Rossmann folds, and Rossmann-like folds (Figs. 6.1 and 6.2).

Once ancient protein folds were established, they were duplicated as separate genes and evolved to form a host of enzymes all built on the same or a similar scaffold.

Over billions of years, many of these scaffolds have not been improved upon or replaced.

Chapter 7

The Best Art

Molecular biology, genetics, and biochemistry have the best science.

I respect others sciences. I just like mine better.

We also have the best art.

I do not think we have the best poetry, song, or music. Perhaps, someday, we may.

To fully appreciate this book, you must learn to make molecular art.

I cannot teach this to you, so you must teach yourself. Actually, one on one, I could teach this to you in about an hour. It will not take you much longer than this using available tutorials, and it will be worth the effort.

First, if I can learn this, so can you. I am an ancient, multiply retooled biochemist with less than adequate computer skills.

This is how you do it.

First, download a protein visualization program. I suggest YASARA, Visual Molecular Dynamics (VMD), Pymol, and/or Chimera. I have a working knowledge of these programs, and you are smarter than me, so this will be easy for you. Run a tutorial or two. There is huge documentation for the use of these programs, which have the complexity of a reasonably simple computer game but with a much greater pay out. Also, this is the perfect computer video game because you will never win no matter how many times you play, so you will remain frustrated, motivated, and unsatisfied your entire life: as is natural for humans.

To feed your molecular graphics program, learn to download protein coordinates as PDB files from the RCSB (Research Collaboratory in Structural Bioinformatics) protein data bank. Learn to scroll through and view the documentation pages for your PDB file.

OK. Now we start.

In 2017, molecular art is the best art made by humans (see Fig. 7.1).

Evolution since Coding. http://dx.doi.org/10.1016/B978-0-12-813033-9.00007-X

FIGURE 7.1 Molecular biology has the best art. A TIM barrel protein. β-sheets are yellow. α-helices are purple. Turn is cyan. Coil is white.

Chapter 8

Intelligent Design

Protein structure/function does not support the concept of intelligent design.

Protein design follows a different set of rules: the rules of gene fusion, evolution, genetics, and protein folding.

As we continue, I shall address a host of religious issues. After all, this is a story of the genesis of life on earth, which is a potentially religious subject. I do not think that evolution or protein structure/function is in any way antireligion, at least, no more antireligion than digital watches (Fig. 8.1) and jet airplanes.

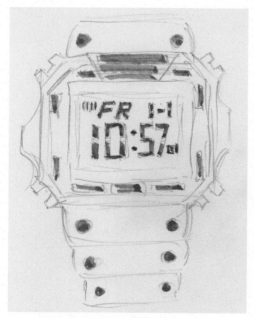

FIGURE 8.1   A digital watch.

Evolution since Coding. http://dx.doi.org/10.1016/B978-0-12-813033-9.00008-1

Chapter 9

Polymers

RNA is a linear polymer of four nucleic acids.

DNA is a linear polymer of four nucleic acids.

Protein is a linear polymer of 20 amino acids.

RNA can fold into a complex three-dimensional shape that may be capable of directing complex chemistry.

RNA enzymes are referred to as "ribozymes."

RNA can also function as genetic code.

RNA-based chemistry can be enriched via chemical modifications of nucleic acid bases.

Life comes from an ancient RNA world and RNA-protein world (~3.5–4.1 billion years ago).

Compared to RNA, DNA tends to function as more chemically inert but more stable double-stranded structure. As such, DNA is a more reliable repository of genetic information than RNA—hence the evolutionary advantage of DNA genomes over more ancient RNA genomes. Currently, all cellular life relies on DNA genomes. Some viruses utilize RNA genomes.

Because of its inaccessibility, low reactivity, and chemical stability, DNA took over from RNA as the primary genetic material.

Proteins are linear polymers of 20 amino acids. Because of the diverse chemistry of the amino acids and because proteins fold into distinct clustered shapes, proteins tend to be more capable catalysts than RNA enzymes (ribozymes), although perhaps not for every possible function.

The activities of many ribozymes were displaced by more able protein enzymes in the RNA-protein world and after.

The ribosome, king of all ribozymes from the RNA-protein world forward, is a notable exception to this trend of ribozyme displacement by protein catalysts. For its role in amino acid polymerization, the ribosome has never been successfully challenged by protein enzymes for its core function in protein synthesis. This being said, however, the peptidyl transferase center of the ribosome is not a very good enzyme to catalyze peptide bond formation. The peptidyl transferase center appears to be mostly a molecular crowding agent to cluster amino acylated tRNAs to support peptide bond synthesis, which then occurs as a spontaneous reaction.

This is a story about polymers and repeating sequences: RNA, DNA, protein, α/β proteins, RIFT barrels, double-Ψ–β-barrels, swapped-hairpin barrels,

Evolution since Coding. http://dx.doi.org/10.1016/B978-0-12-813033-9.00009-3

TFB, σ factors, TBP, HTH (helix-turn-helix) motifs, WHTH (winged helix-turn-helix) motifs, TATA boxes, BREs (TFB-recognition elements), etc.

Life on earth is a story of developing complexity via generation of repeating sequences and motifs.

One of the key concepts of evolution of life on earth is that biological complexity is often built up via (1) iteration of a simple chemical or motif; (2) coevolution with interacting factors and chemicals; and (3) simplification (or not) depending on evolutionary pressures.

Think: polymerization/iteration, before considering intelligent design. Dumb repetition is much more likely. Biological designs do not appear to be intelligent.

Chapter 10

LEGO (Trademark) Life

The story of genesis of life on earth is preserved as written history. The remarkable story remains written in genetic code, protein secondary structure, and protein tertiary structure. This is a language that can be read by anyone (no PhD necessary, although perhaps helpful). The story has now largely been decoded.

In this chapter, the story of genesis is told in its simplest form, while in subsequent chapters, it is told in more detail with prettier pictures.

As related here, the story of life on earth reduces to a few simple slides.

To emphasize the simplicity of the evolutionary models in this book, I offer the following LEGO (Trademark) life version of genesis.

At its most primitive core, life is made up of RNA synthesis, translation (protein synthesis), and metabolism. Modern mechanisms of DNA replication came later, after last universal cellular common ancestor (LUCA) and after divergence of archaea and bacteria. At LUCA, DNA replication was via transcription (RNA synthesis) followed by reverse transcription. In the process of reverse transcription, DNA is synthesized from RNA using a reverse transcriptase enzyme, such as those used for retroviral replication and DNA integration.

In all cellular and DNA genome-based life, ribosomes catalyzed messenger RNA templated "translation" for protein synthesis. Ribosomes or protoribosomes caused the RNA-protein world to displace the RNA world. Until recently, no simple model for genesis of primitive ribosome RNA sequences was available, but I will offer one here. I posit that transfer RNA (tRNA) is the central molecule for evolution of protein synthesis and translation systems. Translation systems can readily be built around tRNA, and the evolution of tRNA is a surprisingly simple and straightforward story. Ribosomes and ribosomal RNA (rRNA) are another murkier story. Because ribosomes are partly ribozymes and because RNA tends to mutate faster than protein, the simplest initial patterns of ribosomal RNAs may never be known. Frankly, it is a big surprise (to me) that so many of the initial and simplest protein patterns remain discernable. The story for tRNA is similarly simple. For genesis of the first ribosomes, I would suggest a model involving many RNA ligations creating large knots of RNA with some functional parts.

Although ribosome and translation may remain somewhat enigmatic, RNA synthesis and metabolism can be described in starkly simple terms using small, protein LEGO (Trademark) pieces. tRNA, the core molecule in translation, can also be described similarly.

Evolution since Coding. http://dx.doi.org/10.1016/B978-0-12-813033-9.00010-X

To generate RNA synthesis, we need RNA polymerases of the two DPBB (two double-Ψ–β-barrel) types, so we need DPBBs (Fig. 10.1). DPBBs are a simple ββαββββαβ fold (a ligated RNA encoding a ββαβ repeat) generated in the RNA-protein world. A simple model for generation of DPBBs is shown. Think of ββαβ as a LEGO (Trademark) piece. Link two ββαβ units together and fold.

RIFT barrel and DPBB enzymes are essential for many core biological processes: i.e., transcription, translation, and replication. RIFT barrels and DPBBs radiate to all cellular life in the three domains: archaea, bacteria, and eukaryotes.

To generate RNA coding of proteins, tRNA is required (Fig. 10.2). Most likely, 2-amino acylated tRNAs and a peptidyl transferase center are minimally required for peptide bond formation, and overly ornate translation systems can be evolved from this core. "Cloverleaf" tRNAs were generated

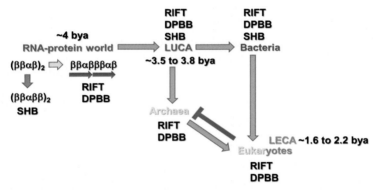

FIGURE 10.1 A simple model for generation of cradle-loop barrels, which arose in the RNA-protein world. *Small yellow arrows* indicate ligations. *Small red arrows* indicate repeats. DPBB for double-Ψ–β-barrels. SHB for swapped-hairpin barrels. Eukaryotes are posited to have competed with their parent archaea (red symbol).

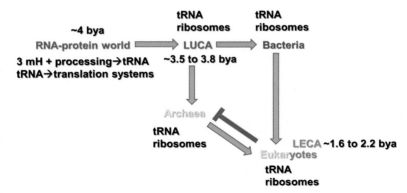

FIGURE 10.2 Evolution of translation systems. Cloverleaf tRNA was an early and radical innovation. Translation systems evolved around tRNA.

from three minihelices ligated together followed by internal RNA processing. Cloverleaf tRNAs, which are now >3.5 billion years old, represent a radical, defining, and very early evolutionary innovation in generation of RNA-template-dependent protein synthesis systems. In many ways, tRNA is the core and central function in molecular coding.

To generate metabolism, we need enzymes and oxidoreductases (redox proteins; battery-powered life). These are obtained from TIM barrels $(\beta-\alpha)_8$ (many enzymes) and Rossmann folds $(\beta-\alpha)_8$ (oxidoreductases) (Fig. 10.3). Many other enzymes, particularly ones utilizing nucleoside substrates (i.e., kinases, GTPases, ATPases), are generated from Rossmann folds. A simple model for evolution of TIM barrels and Rossmann folds is shown. It is likely that smaller $\beta\alpha\beta\alpha$ fragments interacted functionally with RNA in ways that are currently (till 2017) unknown but might yet be discerned.

Utilization of DNA templates, which arise at LUCA, requires promoters and general transcription factors (Fig. 10.4). Promoters are DNA sequences to which RNA polymerase binds and from which RNA polymerases initiate an RNA chain. Promoters are described in detail in Chapter 27. At LUCA, to recognize a promoter DNA sequence, a primordial initiation factor is modeled as a four HTH (helix-turn-helix) factor. Bacterial sigma (σ) factors are modeled as four HTH factors, in which HTH_1 is now vestigial and HTH_2 is specialized for opening the promoter DNA-10 region. Archaeal TFB is a two HTH factor, with archaeal TFB HTH_1 and HTH_2 corresponding to bacterial sigma HTH_3 and HTH_4. In eukaryotes, three archaeal TFB homologues are identified: Rrn7 for RNA polymerase I, TFIIB for RNA polymerase II, and Brf1 for RNA polymerase III. The naming of factors is historic, which may be confusing.

Another simple story relates to evolution of "transcription" (RNA synthesis; catalyzed by DDRPs: DNA-dependent RNA polymerases) and "replication" (DNA synthesis; catalyzed by DDDPs: DNA-dependent DNA polymerases)

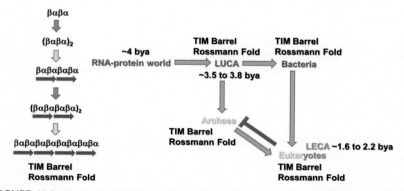

FIGURE 10.3   A simple model for evolution of metabolism. TIM barrels and Rossmann folds (posited to be a rearranged TIM barrel) are posited to have arisen in the RNA-protein world and to have radiated to all subsequent life on earth.

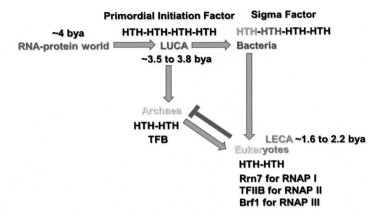

FIGURE 10.4   A simple model for evolution of a core general transcription factor. A four HTH primordial initiation factor is posited to have been present at LUCA and radiated to form sigma factors in bacteria (four HTH), TFB in archaea (two HTH) and Rrn7, TFIIB and Brf1 in eukaryotes (two HTH).

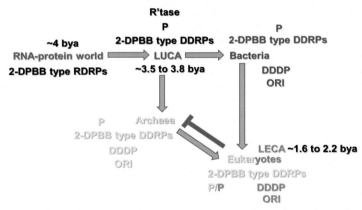

FIGURE 10.5   Evolution and utilization of RNA and DNA templates. *DDDP*, DNA-template-dependent DNA polymerase; *DDRP*, DNA-template-dependent RNA polymerase; *DPBB*, double-Ψ–β-barrel; *ORI*, origin of replication; *P*, promoter; *RDRP*, RNA-template-dependent RNA polymerase; *R'tase*, reverse transcriptase. Colors represent bacterial (blue), archaeal (yellow), and eukaryotic (pink) innovations.

systems (Fig. 10.5). Transcription by RNA polymerases launches from promoter (P) DNA sequences. Replication launches from replication origins (ORIs). This simple story is told here to emphasize the probable time line for genesis and the probable replication system at LUCA. Notably, separate replication systems (DDDPs: DNA-dependent DNA polymerases) and replication origins (ORIs) are posited to evolve after divergence of archaea and bacteria. This means that, at LUCA, the mechanism of DNA replication is the mechanism used by retroviruses today: transcription by a DDRP from a promoter followed by reverse transcription by a reverse transcriptase. As the story of genesis unfolds, the reader may wish to revisit this figure. This story is important because it explains the core relationship between RNA templates, DNA templates, and proteins.

As shown in Fig. 10.5, the ancient RNA-protein world evolves to LUCA, which is a DNA genome-based organism(s). LUCA diverges to bacteria and archaea very early in evolution. At LECA (the last eukaryotic common ancestor), a α-proteobacterium invades a Lokiarchaeota archaea, giving rise to eukaryotes. The α-proteobacterial endosymbiont becomes the mitochondria. A selfish DNA element called the group II intron of the α-proteobacteria attacks the Lokiarchaeota genome, creating introns within genes and forcing evolution of the cell nucleus to minimize translation of introns on ribosomes. Many eukaryotic innovations are forced by the catastrophic fusion of bacteria and archaea and the unleashing of α-proteobacterial group II introns (more about this later).

There are multiple key initial concepts to take from this figure. Transition from the RNA-protein world to the DNA genome world (LUCA) requires two double-Ψ–β-barrel-type RDRPs (RNA-template-dependent RNA polymerases) to become DDRPs (DNA-template-dependent RNA polymerases). DDRPs initiate transcription from promoter (P) DNA sequences, which must evolve at LUCA. At LUCA, the mechanism of DNA replication is transcription by DDRPs from a promoter DNA sequence followed by reverse transcription by a reverse transcriptase. Separate DDDPs (DNA-template-dependent DNA polymerases) and origins of replication (ORIs) evolve separately in bacteria and archaea after divergence. Eukaryotic transcription and replication systems resemble archaeal systems with many eukaryotic innovations. As we continue, these stories are told in detail. Here, the story is told in outline, but, when the figures in this chapter are mastered, the reader understands genesis of life on earth. In concept, this is the story.

At the risk of mixing metaphors, evolution of life on earth can be considered in terms of (1) LEGO (Trademark) life; and/or (2) cutout paper dolls (describing β sheet interactions in barrels and sheets). The core stories of genesis are starkly simple. The code has largely been broken, and, after ~4 billion years, the code remains surprisingly easy to read, at least for RNA synthesis, DNA synthesis, tRNA and protein synthesis, and metabolism.

Core protein motifs have an ancient history that can be traced back to LUCA (>3.5 bya) or to the RNA-protein world (~4 bya).

A few years ago (i.e., 2013), I could not imagine these models. Now, I consider them to be among the strongest and most likely ideas in the modern biological sciences. These models render an otherwise incomprehensible story of genesis simple.

This chapter on LEGO (Trademark) life represents the entire book in an outline form.

Chapter 11

α/β Proteins

I wish for you, with your newfound, immense skill in protein structure analysis, to consider α/β proteins (Fig. 11.1).

We shall consider TIM barrels, Rossmann folds, TOPRIM domains, ATPases, and kinases. Many more examples are available than we shall cover.

Feel free to follow along and check my work (i.e., using Pymol, VMD, Chimera, and/or YASARA). In case I make a mistake, you can correct me.

Even a layman can do this, and, in science, simple is good.

There is a classic text by Branden and Tooze on protein structure that was an inspiration to me thinking about α/β proteins. Initially, I was somewhat intimidated by this book, but, after gaining some insight into β-α repeat proteins, the book made perfect sense.

So, what is the take home lesson?

α/β proteins are simple alternating β-sheet and α-helix repeats (Fig. 11.1), and they have been such since last universal cellular common ancestor (LUCA) (~3.5 billion years) and the RNA-protein world (~4 billion years). α/β proteins remain recognizable as β-α repeats today. As long as there has been genetic coding, proteins, and fundamental metabolism, α/β proteins existed as immortal repeating designs. If this does not thrill and chill you, you are too serene. You gaze on immortality. As far as you and I are concerned, 4 billion years is immortal.

An additional insight that arises from recognizing the recurring pattern in 4 billion-year-old β-α repeat proteins is this: much of protein biochemistry, evolution, and genetics reduce to a simple iterated pattern recognition problem. If you can recognize parallel β-sheets, you solve a central and ancient issue in the evolution of life on earth.

Remarkably, α/β repeat folds comprise about 25% of extant proteins in all of life. These are among the most ancient protein folds and the folds with the most parallel β-sheets. FYI: β-sheets can be arranged relative to a neighboring β-sheet either in a parallel or an antiparallel orientation (Fig. 11.1). Because β-sheets form by interaction to another β-sheet, they must exist at least in pairs. So, by itself, a β-sheet is not a β-sheet. Without a partner, a β-sheet could fold into a α-helix or something even worse. TOPRIM domains are $\sim(β\text{-}α)_{4-5}$ repeats [i.e., $(α\text{-})β\text{-}α\text{-}β\text{-}α\text{-}β\text{-}α\text{-}β\text{-}α$], Rossmann folds are $(β\text{-}α)_8$ repeats (i.e., β-α–β-α–β-α–β-α–β-α–β-α–β-α–β-α), and TIM barrels are $(β\text{-}α)_8$ repeats (β-α–β-α–β-α–β-α–β-α–β-α–β-α–β-α). TOPRIM domains and Rossmann folds are twisted linear sheets. TIM barrels dance in a circle to close the barrel. It is likely that

Evolution since Coding. http://dx.doi.org/10.1016/B978-0-12-813033-9.00011-1

33

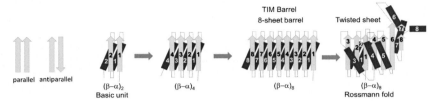

FIGURE 11.1   TIM barrels and Rossmann folds comprise some of the oldest and most enduring enzymes. TIM barrels and Rossmann folds are generated from repetition of a β-α–β-α unit. β-sheets (*yellow arrows*) and α-helices (*purple rectangles*) are indicated. β-sheets can be either parallel or antiparallel. In (β-α)$_n$ repeat proteins, the primary fold includes only parallel β-sheets.

a Rossmann fold is a rearranged TIM barrel. TOPRIM domains were likely evolved from degeneration of a Rossmann fold (β-α)$_8$ → ~(β-α)$_{4–5}$.

If the reader has suffered through beginning biochemistry, he/she may know that TOPRIM stands for topoisomerase-primase, two enzymes required in DNA genome maintenance. Without TOPRIM domains, there could be no DNA genomes. Without Rossmann folds there could be no citric acid or glyoxylate cycle. There would be no ancient redox energy transduction: which is battery-powered life. Without TIM barrels (TIM for triose phosphate isomerase) there would be no glycolysis or carbon dioxide fixation by cyanobacteria, algae, or plants. Adding some more complex α/β folds, you get most of the remaining core functions of life, including kinases, ATPases, and GTPases, so chemical signaling and energy transduction are also features of α/β repeats. Certainly beginning biochemistry would be mortally wounded without α/β proteins (which sadly, in teaching, are much too seldom identified as such). Teaching biochemistry would be so much more interesting if the story of α/β repeat proteins were told. An absence of α/β proteins would destroy life on earth, because α/β proteins support ancient metabolism and redox and chemical energy transduction. I am not certain if I have been clear, but I believe I just brought unprecedented concept to the study and teaching of biochemistry. Viewed from the perspective of ancient evolution and generation of β-α repeat proteins, biochemistry is a highly conceptual and a frighteningly simple subject. We may persist in teaching biochemistry as stamp collecting, but evolution provides meaning and simplicity to biochemistry.

Because α/β proteins can be referred to as (β-α)$_n$ repeats, this may seem confusing. Once iteration of the basic motif (β-α)$_2$ occurs, of course, it does not matter whether α or β comes first or second, because, depending on the start, a (α–β) repeat can also be considered a (β-α) repeat, and an α- or β-unit can be lost at either ends of the chain by genetic deletion. This being said, however, inspection of many TIM barrels and Rossmann folds indicates that the original pattern is β-α–β-α, and that the β-sheet comes first. Furthermore, there is the issue of the paired β-sheets. The elementary unit for (β-α)$_n$ repeats appears to be (β-α)$_2$ not (β-α)$_1$. The likely explanation is that, unless within a dimer or a larger structure, (β-α)$_1$ cannot form a β-sheet because a β-sheet requires

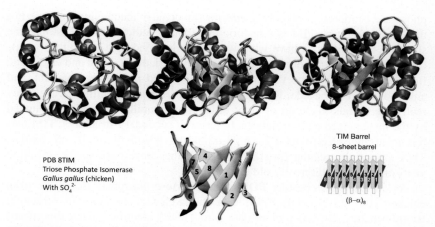

FIGURE 11.2 A TIM barrel protein (PDB 8TIM): *Gallus gallus* (chicken) triose phosphate isomerase. Two views. SO_4^{2-} is a phosphate $\left(PO_4^{2-}\right)$ mimic to indicate the enzyme active site. So far as I know, active sites of TIM barrel enzymes always locate to the C-terminal end of the β-sheets. β-sheets (*yellow arrows*) and α-helices (*purple rectangles*) are indicated.

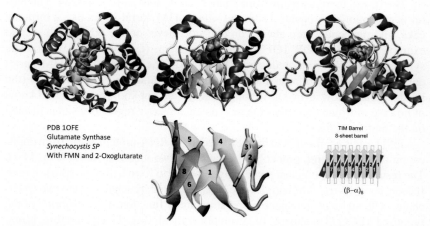

FIGURE 11.3 A TIM barrel protein (PDB 1OFE): *Synechocystis* Sp. glutamate synthase, a cyanobacterial FMN-dependent oxidoreductase. Three views. FMN and 2-oxoglutarate are shown to indicate the enzyme active site. β-sheets (*yellow arrows*) and α-helices (*purple rectangles*) are indicated.

another β-sheet as a partner. (β-α)$_2$, by contrast, can form two parallel interacting β-sheets, and the basic polymerization unit is (β-α)$_2$ (Fig. 11.1).

Let us start with TIM barrels (β-α)$_8$ (Figs. 11.2 and 11.3). The reason for starting with TIM barrels is that Rossmann folds (β-α)$_8$ can be considered to be protein origami rendered from rearranged TIM barrels. TOPRIM domains ~(β-α)$_{4-5}$ are likely derived from Rossmann folds (β-α)$_8$. TIM stands for triose phosphate isomerase, a glycolytic enzyme (involved in metabolizing sugar). For the barrel structure, TIM is a historic designation indicating the first enzyme structure discovered to have what was later identified as a ubiquitous $(\beta\text{-}\alpha)_8$ barrel fold. TIM barrel enzymes provide many, varied metabolic and enzymatic functions throughout the vast tree of life dispersed in archaea, bacteria, and eukaryotes. From the side, TIM barrel structures resemble a birthday cake. They really are barrels made up of eight ordered and parallel β-sheets surrounded by the interspersed eight α-helices. In terms of secondary structures, TIM barrels are arranged as $(\beta1\text{-}\alpha1\text{–}\beta2\text{-}\alpha2\text{–}\beta3\text{-}\alpha3\text{–}\beta4\text{-}\alpha4\text{–}\beta5\text{-}\alpha5\text{–}\beta6\text{-}\alpha6\text{–}\beta7\text{-}\alpha7\text{–}\beta8\text{-}\alpha8)$. They are intially regular $(\beta\text{-}\alpha)_8$ repeats, and they have been so for ~4 billion years on an ~4.6 billion-year-old earth.

As an example, consider PDB 8TIM, a triose phosphate isomerase from chicken (Fig. 11.2). PDB 8TIM is a Research Collaboratory in Structural Bioinformatics (RCSB) protein data bank (PDB) identifier. What this means is that if you go to the RCSB PDB website, you can download the atomic coordinates for this or any other protein in which you might have an interest. Furthermore, you can spin the molecule around and around on a personal computer screen until you are dizzy.

The $(\beta\text{-}\alpha)_8$ barrel extends from phenylalanine-6 to isoleucine-244. The barrel is exceptionally regular. From the side, a TIM barrel looks like a birthday cake, and the candles are always on one side, never on the other. As a second example, consider PDB 1OFE *Synechocystis* Sp. (cyanobacterial) glutamate synthase, an FMN oxidoreductase (Fig. 11.3). The active site of TIM barrel enzymes is always (so far as I know) at the C-terminal end of the β-sheets (the heads of the yellow arrows). Although, sadly, I do not know everything, I know of no case in which the active site or even an allosteric regulatory site is located on the bottom of the cake.

TIM barrels are thought to have evolved in the RNA-protein world from duplication of a $(\beta\text{-}\alpha)_2$ unit, probably by ligation of two identical RNAs (Fig. 11.1). The resulting $(\beta\text{-}\alpha)_4$ unit can potentially dimerize to form a dimeric $(\beta\text{-}\alpha)_8$ barrel. Indeed, some $(\beta\text{-}\alpha)_8$ TIM barrels have been separated by scientists into two $(\beta\text{-}\alpha)_4$ chains without huge damage to barrel structure or function. The feasibility of the proposed LEGO (Trademark) life evolutionary path (Fig. 11.1), therefore, is reasonably demonstrated by experiment. A second RNA ligation can link two $(\beta\text{-}\alpha)_4$ chains into the $(\beta\text{-}\alpha)_8$ barrel. Because protein LEGO (Trademark) life evolution occurs in a rich RNA background, it will be fascinating to analyze interactions of protein LEGO bits (i.e., $(\beta\text{-}\alpha)_2$ and $(\beta\text{-}\alpha)_4$) with RNA. For instance, $(\beta\text{-}\alpha)_2$ units may initially be evolved for functional interactions with single-stranded or double-stranded RNA, because, in an evolutionary scheme, $(\beta\text{-}\alpha)_2$ units cannot have foreknowledge of becoming functional $(\beta\text{-}\alpha)_4$ or $(\beta\text{-}\alpha)_8$ units. During evolution, proteins can diverge to assume new functions.

In triose phosphate isomerase, the TIM barrel is often a slightly squashed barrel, probably to exclude water. As we shall see, some of the most compact β-barrels are six antiparallel β-sheet barrels. There must be a reason that, to form a barrel with all parallel β-sheets, eight parallel β-sheets is generally favored over six or seven parallel β-sheets. Perhaps this relates to the proposed initial ligation of two $(\beta\text{-}\alpha)_4$ units to form the $(\beta\text{-}\alpha)_8$ barrel. Part of the driving force for the barrel is the natural curvature of the linear array of β-α repeats. The overall shape of the TIM barrel is similar to a human red blood cell (a puckered toroidal shape lacking a hole). The enzyme active site, therefore, sits in a kind of slightly sheltered funnel-shaped cavity.

If for some reason you do not believe me that TIM barrels are ubiquitous, consider NCBI Conserved Domain Database cl21457: TIM_phosphate_binding_superfamily. You will find an immense collection of molecular birthday cakes with the candles always on the top (the C-terminal end of the β-sheets). Similar to RNA, DNA, and proteins, TIM barrels are dumb $(\beta\text{-}\alpha)_8$ repeats and have been such forever (~4 billion years is forever to a human).

Rossmann folds are a very large and important class of ancient $(\beta\text{-}\alpha)_8$ repeat proteins (i.e., Fig. 11.4). The Rossmann fold forms as a twisted linear β-α repeat. Ancient metabolism to generate energy is based either on redox (oxidation-reduction) chemistry or chemical energy, i.e., ATPases and GTPases. NAD and FAD oxidoreductases (redox proteins) are Rossmann fold proteins. Veterans of biochemistry courses will recognize these important enzymes as the dominant constituents of the citric acid cycle and the glyoxylate cycle that one was forced to memorize and through which one was forced to trace (on paper) radioactive carbon atoms. For the purposes of this book (2017), Rossmann folds will be considered a relic of TIM barrels or a closely related $(\beta\text{-}\alpha)_8$ repeat. Basically, compared to a TIM barrel, Rossmann folds are proposed to be a twice

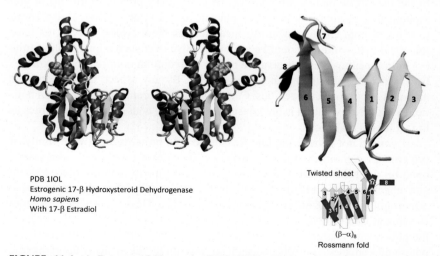

PDB 1IOL
Estrogenic 17-β Hydroxysteroid Dehydrogenase
Homo sapiens
With 17-β Estradiol

Twisted sheet

$(\beta\text{-}\alpha)_8$
Rossmann fold

FIGURE 11.4 A Rossmann-fold protein (PDB 1IOL): *Homo sapiens* estrogenic 17-beta-hydroxysteroid dehydrogenase. Estradiol indicates the location of the active site, at the C-terminal end of the β-sheets. The Rossmann fold appears to be a rearranged TIM barrel (see Fig. 11.1 for a model). β-sheets (*yellow arrows*) and α-helices (*purple rectangles*) are indicated.

rearranged linear twisted sheet. The initial rearrangement to form a Rossmann fold is to place β4 next to β1 rather than next to β3, as in a TIM barrel (Fig. 11.4). The second rearrangement from the TIM barrel to form a Rossmann fold is to pair both β7 and β8 with an elongated β6. Because loops and α-helices separate each β-sheet, there is sufficient flexibility in the structure to make the rearranged folds. Rearrangement in the Rossmann fold is necessary to prevent curvature into a horseshoe or barrel shape. The more linear Rossmann sheet is maintained by changing the orientation and curvature of β1–β3 relative to β4–β8. Essentially, the more linear arrangement is maintained because curvature of β1–β3 and β4–β8 tends in two opposing directions. A potential protein origami model for generation of a Rossmann fold from a TIM barrel is shown in Fig. 11.1.

Homo sapiens estrogenic 17-beta-hydroxysteroid dehydrogenase (PDB 1IOL) is a typical Rossmann fold $(\beta\text{-}\alpha)_8$ linear, twisted sheet protein (Fig. 11.4). As in TIM barrels, the business end of the molecule is stacked at the C-terminal ends of the eight parallel β-sheets. The Rossmann fold was named after Michael Rossmann (Purdue University), an eminent X-ray crystallographer. In the figure, secondary structure elements are numbered (β1–α1–β2–α2–β3–α3–β4–α4–β5–α5–β6–α6–β7–α7–β8–α8). The protein is a remarkably consistent $(\beta\text{-}\alpha)_8$ repeat. As a second example, *Brassica napus* β-keto-acyl carrier protein reductase, a plant-NADP-dependent oxidoreductase, is shown (PDB 1EDO) (Fig. 11.5). In this case, α8 has been deleted from the C-terminus of the protein. Although some Rossmann fold–derived protein folds are shortened to $(\beta\text{-}\alpha)_{4\text{-}7}$, astoundingly, in ~4 billion years, for many Rossmann folds there has been little change in the overall secondary structure of the core $(\beta\text{-}\alpha)_8$ protein motif. So, if you thought that evolution is infinitely innovative, you were wrong. I also admit to being wrong. Until I began to concentrate on these issues, I could not have imagined that evolution could be so conservative over such a long span of time (~4 billion years). The conservative nature of evolution requires some explanation, which I shall attempt in a later chapter. By any aspect, the conservative nature of evolution is miraculous. Certainly, in 2017, comparing American politics to real life, evolution is not considered to be so "conservative" in politics as in fact it is in real life.

If you do not believe me that Rossmann fold proteins are a vast family of proteins throughout the three domains of life (archaea, bacteria, eukaryotes), open up the NCBI (National Center for Bioinformatics Information) website and check out the cI21454: SDR (short-chain dehydrogenases/reductases) Superfamily in the Conserved Domain Database. There, you will find more Rossmann-fold protein structures and sequences than you (or I) will ever have the patience to analyze. You can load any of the many PDB files to your computer and convince yourself that they are essentially the same as the reductase shown in overall fold and secondary structure. Rossmann-fold pillars of life are all derived from $(\beta\text{-}\alpha)_8$ repeats and have been forever (to a human ~4 billion years is forever).

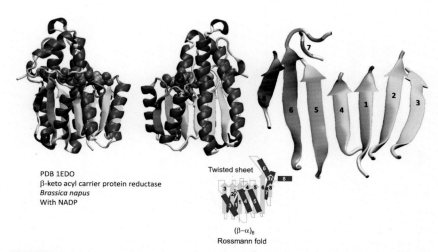

PDB 1EDO
β-keto acyl carrier protein reductase
Brassica napus
With NADP

Twisted sheet

(β-α)₈
Rossmann fold

FIGURE 11.5   A Rossmann-fold protein (PDB 1EDO): β-keto-acyl carrier protein reductase, a plant NADP-dependent oxidoreductase. NADP is located at the active site, at the C-terminal end of the β-sheets.

Similar to Rossmann folds, TOPRIM domains ~$(\beta\text{-}\alpha)_{4-5}$ are another linear twisted sheet that appears to be generated by degradation of a larger repeat (Figs. 11.6 and 11.7). Download PDB 2RGR from the RCSB protein data bank and load it into a molecular structure visualization program on a personal computer. I like 2RGR (Fig. 11.6) because it has a Mg^{2+} ion associated with the TOPRIM domain, and many TOPRIM X-ray crystal structures (i.e., 1T6T; Fig. 11.7) sadly lack Mg^{2+}, because Mg^{2+} was omitted by researchers during crystallization. TOPRIM domains evolved as an Mg^{2+} chelator to participate in a large number of protein–nucleic acid interactions. In the RNA-protein world, a $(\beta\text{-}\alpha)_4$ polymer fold was present as a precursor to TIM barrels, but extant TOPRIM domains are cut from a larger $(\beta\text{-}\alpha)_n$ repeat ($n > 4$), as I shall explain. TOPRIM domain proteins that are necessary to open the DNA double helix were certainly required with advent of DNA genomes at about the time of LUCA.

The 2RGR structure is of a yeast DNA topoisomerase: DNA gyrase type, ATP-dependent, Type IIA DNA topoisomerase (Fig. 11.6). This enzyme generates negative supercoils in DNA, and underwinding DNA, which facilitates helix opening, is necessary for transcription, replication, and recombination. The TOPRIM domain is located from amino acid threonine 444 to phenylalanine 559. If you draw in the Mg^{2+} it will help your understanding. The Mg^{2+} is located very close to the approximate pseudosymmetry axis of the TOPRIM domain. Mg^{2+} is held at the carboxy terminal end of the two central β-sheets, between β1 and β3. All of the β-sheets point in the same orientation (N→C; parallel) toward Mg^{2+}. Gazing at the 2RGR sequence to glutamine 656, the TOPRIM domain looks as if it was folded from a $(\beta\text{-}\alpha)_n$ repeat ($n \geq 6$) (not shown in image). Another TOPRIM domain structure is 1T6T. The TOPRIM

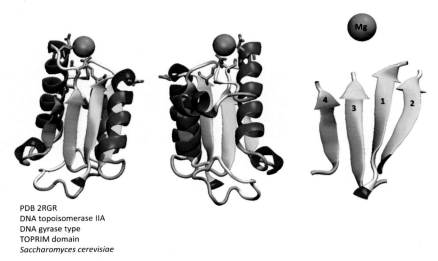

PDB 2RGR
DNA topoisomerase IIA
DNA gyrase type
TOPRIM domain
Saccharomyces cerevisiae

FIGURE 11.6   A TOPRIM domain. PDB 2RGR is shown. β-sheets (*yellow arrows*) and α-helices (*purple rectangles*) are indicated. The chelated Mg^{2+} forms part of the topoisomerase active site. The TOPRIM fold appears to be a simplified version of a Rossmann fold.

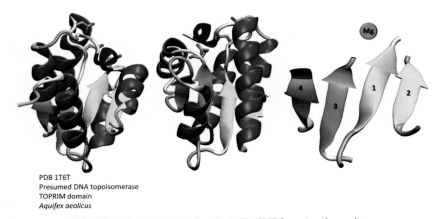

PDB 1T6T
Presumed DNA topoisomerase
TOPRIM domain
Aquifex aeolicus

FIGURE 11.7   A TOPRIM domain. PDB 1T6T from *Aquifex aeolicus*.

domain extends between valine 23 and lysine 103. Because this structure has an extra α-helix at both ends, this TOPRIM domain was also folded from a longer $(β-α)_n$ repeat.

So far, using the complexity of simple cutout dolls or LEGOs (Trademark), we have constructed glycolysis (TIM barrels), the citric acid cycle (Rossmann folds), and the glyoxylate cycle (Rossmann folds) (Fig. 11.1). We have also constructed essential DNA-protein functions (TOPRIM domains). To make a living system, however, you need more. Redox chemistry (Rossmann folds and

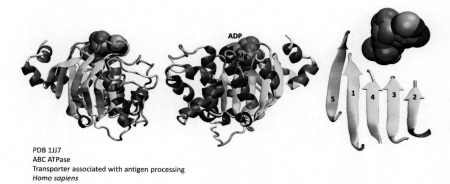

PDB 1JJ7
ABC ATPase
Transporter associated with antigen processing
Homo sapiens

FIGURE 11.8   An ABC ATPase protein (PDB 1JJ7): human ABC ATPase transporter associated with antigen processing. ADP is shown to indicate the active site. The rearranged Rossmann-derived part of the fold is highlighted in the rightmost image.

some TIM barrels) is like running on battery power. You also need chemical energy transduction (ATPases, GTPases) and cellular modification and communication (kinases). These functions are also dominated by α/β proteins derived from $(\beta\text{-}\alpha)_n$ repeats. Examples are shown in Figs. 11.8 and 11.9. Fig. 11.8 shows an image of complex α/β protein PDB 1JJ7: human ABC ATPase transporter. In the NCBI Conserved Domain Database this staggering family of ATPases is listed as cd00267: ABC_ATPase. Much of this protein is based on a $(\beta\text{-}\alpha)_n$ repeat with some extra genetic graffiti to make a linear, rearranged Rossmann-like sheet into a squashed barrel. The $(\beta\text{-}\alpha)_5$ rearranged Rossmann-like sheet is emphasized in the rightmost image. This is an ancient fold.

In Fig. 11.9, a Rossmann-like fold archaeal tRNA kinase is shown (PDB 3A4L). It appears that this enzyme is derived from a truncated Rossmann fold ($\beta1\text{-}\alpha1\text{-}\beta2\text{-}\alpha2\text{-}\beta3\text{-}\alpha3\text{-}\beta4\text{-}\alpha4\text{-}\beta5\text{-}\alpha5\text{-}\beta6\text{-}\alpha6$) by: (1) reordering of β2 and β3; and (2) deletion or secondary structure rearrangement of β5. The rearrangement of β2 and β3 results in reordering the two sheets during protein folding (β2 and β3 trade places with each other in the linear, twisted sheet).

Figs. 11.10 and 11.11 show two α/β repeat domains from a Swi-Snf ATPase called RapA. This protein is involved in modulating RNA synthesis by bacterial RNA polymerase. Both domains are derived from β-α–β-α–β-α–β-α–β-α–β-α repeats, with different arrangements of the β-sheets. As is also shown in Fig. 11.9, rearrangements of β-sheets are a fairly common event in evolution of new protein folds. Other examples of swapping β-sheets in evolution will be described in other chapters.

So, in summary, α/β repeat proteins (~25% of all proteins) are among the most ancient protein folds known, dating to the RNA-protein world, and, strangely, after ~4 billion years of evolution, they remain recognizable as $(\beta\text{-}\alpha)_n$ repeats. Most protein structures with abundant parallel β-sheets are α/β folds (i.e., TIM barrels, Rossmann folds, TOPRIM domains, ATPases, GTPases,

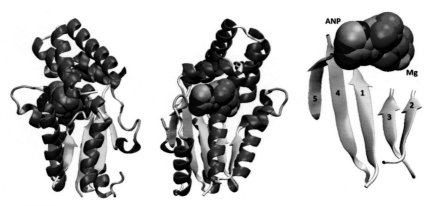

PDB 3A4L
Archaeal O-phosphoseryl-tRNA(Sec) kinase
Methanocaldococcus jannaschii

FIGURE 11.9 A rearranged Rossmann-derived fold protein $(\beta\text{-}\alpha)_5$ (PDB 3A4L): *Methanocaldococcus jannaschii* (archaeal) *O*-phosphoseryl-tRNA kinase. Three views. ANP is a nonreactive ATP mimic to indicate the active site.

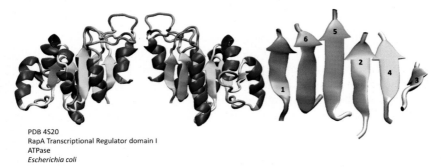

PDB 4S20
RapA Transcriptional Regulator domain I
ATPase
Escherichia coli

FIGURE 11.10 An ancient $\sim(\beta\text{-}\alpha)_6$ domain from a Swi-Snf ATPase RapA.

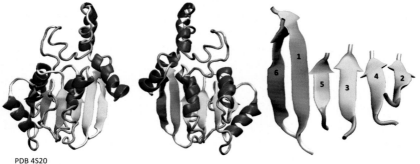

PDB 4S20
RapA Transcriptional Regulator domain II
ATPase
Escherichia coli

FIGURE 11.11 A second ancient $\sim(\beta\text{-}\alpha)_6$ domain from a Swi-Snf ATPase RapA.

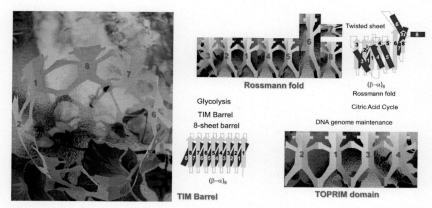

FIGURE 11.12   Ancient $(\beta\text{-}\alpha)_n$ repeat proteins modeled as cutout dolls with β-sheets numbered according to their order in the polypeptide chain.

kinases, and others). Most of the basic biochemistry, metabolism, and energy transduction (redox and chemical), therefore, are based on a small number of ancient α/β folds. Ancient metabolism was in-place in the RNA-protein world (~4 billion years ago) prior to LUCA, the last universal common cellular ancestor of archaea, bacteria, and eukaryotes, ~3.5 to 3.8 billion years ago. If this conclusion does not strain your imagination, you are far too complacent. A few years before writing this (2017), I could not have believed or conceived that this was true, and now I am certain of it.

Ancient evolution is a cutout doll problem generated by repetition of simple and ubiquitous core protein motifs such as β-α–β-α (Fig. 11.12). Life is made of and evolved from genetic LEGO (Trademark) pieces.

Chapter 12

The Inevitability of α/β Folds

Stephen J. Gould, eminent evolutionary biologist, indicated that if evolution were to occur again, it would look quite different. Maybe so, particularly for intact animals. I heard Gould speak at a Michigan State University commencement.

With regard to α/β $(\beta-\alpha)_n$ repeat proteins, however, the evolutionary path of life on earth seems to have been almost assured, if not inevitable, and I can imagine it happening again and again in much the same fashion.

Why do I think this?

Given rules of genetics and the genetic code and rules for protein structure, solubility, and function, it is very difficult to conceive of life on earth without a primary role for $(\beta-\alpha)_n$ repeat proteins.

Currently ~25% of all known proteins are α/β folds, and most protein folds are known.

So, why α/β folds?

$(\beta-\alpha)_n$ repeat folds were one of the fastest routes to evolve to stable structure and solubility.

At some level, ancient evolution is a race, and the fastest to fold may be the fold to win. Once workable protein folds are established, they are seldom replaced.

Ignoring turns and coils, protein secondary structures are made up of β-sheets and α-helices.

In order to form a structured protein, β-sheets are very important because they link together with ordered and directional hydrogen bonds. Interacting β-sheets, therefore, give structure to proteins.

β-sheets, however, can cause aggregation of proteins similar to the amyloid plaques in Alzheimer's disease that form by inappropriate associations of β-sheets. So, without accurate folding, too much β-sheet in a protein may cause solubility problems.

α-helices, however, help to solubilize β-sheets.

One of the fastest means to generate stable and soluble protein structures using genetics, therefore, appears to be through $(\beta-\alpha)_n$ repeats. As I have described, genetic duplication and oligomerization mechanisms are common. $(\beta-\alpha)_n$ repeats are one of the simplest units with both β-sheets and α-helices. Therefore, $(\beta-\alpha)_n$ polymers were among the first structured and soluble protein folds to form. If we replayed evolution, $(\beta-\alpha)_n$ polymers would become a major fold again. If you change the constellation of amino acids so that you get new secondary structures, of course, proteins would look very different.

Evolution since Coding. http://dx.doi.org/10.1016/B978-0-12-813033-9.00012-3

At some level, it is simply an observation that $(\beta-\alpha)_n$ repeat folds are immortal. I assert that these evolutionary designs cannot be perfect, but, also, given evolutionary constraints, they cannot be replaced. Therefore, there are powerful coevolutionary forces on ancient metabolism and ancient enzymes that cannot be overcome in ~4 billion years. This is a remarkable and humbling observation. As I write, I do not fully understand this, and I find it difficult to rationalize. I thought evolution was more innovative.

So, what else might be useful to say about $(\beta-\alpha)_n$ repeat proteins?

β-sheets give structure. α-helices give solubility.

In $(\beta-\alpha)_n$ repeat proteins, active sites (the business end) are always (so far as I know) at the C-terminal end of the β-sheets [in molecular graphics, the points of the red (YASARA), yellow (VMD), or yellow (Pymol) arrows], never at the N-terminal end. Is this observation historic? Based on protein chemistry, I see no compelling reason why active sites are always at the C-terminal end of $(\beta-\alpha)_n$ repeat proteins.

The careful reader knows that I overlook something. In between the β-sheet and the α-helix is turn or coil. I have overlooked turns and coils (loops). So a $\beta-\alpha$ monomer unit is loop-β-loop-α-loop. So, these are not truly simple binary units, as I have described them. Active sites often form at the interface between structure and chaos, because catalysis requires constrained chaos (dynamics). Protein structure and protein pockets allow control over hydration where substrates bind enzymes. Constraining water, amino acid polarizability, dehydration at a protein surface, and the generation of protons and base from water are often key features of enzymatic mechanisms.

Without certainty but with a sense of wonder, I believe that active sites locate to the C-terminal ends of β-sheets because this is how they originally evolved, and they have not changed. This is another indication of the awesome power of coevolutionary forces on these ancient α/β proteins. In 4 billion years of evolution, why cannot a TIM barrel be flipped so that the active site is at the N-terminal end of the β-sheets instead of the C-terminal end? That this has not happened is remarkable. The TIM barrel fold has duplicated and duplicated and metastasized into a myriad of enzyme specificities without (so far as I know) developing an active site or allosteric site at the N-terminal end of the β-sheets. The same for the Rossmann fold.

If someone knows why this might be so, please tell me. If this molecular decision was historic, it was made ~4 billion years ago on an ~4.6 billion year old planet. Remarkably, this is a history written in genetic code and protein secondary structure, so this is a recorded history written long before humans set stylus to papyrus.

I am also curious to know whether one can reengineer a TIM barrel or Rossmann fold to have an active site at the N-terminal ends of the β-sheets.

Further, there is an issue of protein stability relating to β-sheet clusters.

β-sheets that misfold can form insoluble amyloids.

β-sheets are compatible with many amino acid sequences, so potentially β-sheets could be promiscuous and pair with the wrong partner during protein folding, as in the rearrangements of β-sheet folds seen in Chapter 11.

Therefore, to limit capricious folding, the length and stability of β-sheets must generally be limited in evolution.

This is also an evolutionary plea for chaparonin proteins that limit and reverse protein misfolding. Chaparonin proteins can give proteins multiple tries to fold. But chaparonin proteins are not a cure-all and must not be overused in a biological system. Responsibility for proper folding often lies with each enzyme and protein.

So, in addition to evolving functional active sites to support a myriad of substrates, reactions, and specificities, enzymes must evolve to pack near the limits of stability. If an enzyme is too stable, it is in danger of misfolding into insoluble arrays like amyloids.

In conclusion, ancient evolution is a cutout doll problem using the simplest possible protein motifs to build structured and soluble polymers.

It may not have been a deity, therefore, who created life on earth. Rather, it might have been one of god's little children playing with cutout dolls, and life may be more of an accident of chance. If you are committed, therefore, to intelligent design, you must reconcile your views with 25% of life composed of $(\beta-\alpha)_n$ repeats, and you must consider dumb iteration design: evolutionary design.

It is not that a α–β–β unit cannot work in evolution. It is just that, given genetic mechanisms, it is slower to evolve to structure than a β–α–β–α unit.

$(\beta-\alpha)_n$ repeats were inevitable.

Chapter 13

Evolution Puts the Y (Why) in Biology

OK. So it is a dumb pun, even dumber written than spoken, but that is my favorite kind: dumber the better.

It has been said that nothing in biology makes sense except in the light of evolution.

In studying ancient evolution, think of genetic mechanisms: ligations (linking two RNAs), duplications, multimerizations, purifying (negative) selection.

Think of a race to establish structure, solubility, and stability (but not too much stability). Think of establishing dynamic interfaces between structure and chaos. Think $(\beta-\alpha)_n$ repeats. Think cutout dolls. Think LEGO (Trademark) life. Think exceptionally powerful but enigmatic coevolutionary forces. Think chaparonins. Think what could possibly go wrong? Evolution comes down exceptionally hard on unsuccessful designs, because evolutionary editing via natural selection is fierce and unforgiving.

Also, successful core protein folds are a major currency in ancient evolution. You can trace evolution via organisms, genomes, genes, and core protein motifs. Because of a host of chaotic genetic mechanisms (i.e., RNA ligations, genome duplications, endosymbiosis, horizontal gene transfer, gene duplication, jumping genes, multimerization, replication errors, ultraviolet light, radiation), all of these units have currency.

Evolution since Coding. http://dx.doi.org/10.1016/B978-0-12-813033-9.00013-5

Chapter 14

Evolution Is Not Anti-Religion

I do not understand the hatred of evolution by some of the devout.

Evolution is no more antireligion or anti-Bible or anti-Islam than jet airplanes or wristwatches.

Evolution is simply a modern understanding of biology.

Without evolution, nothing in biology makes sense, and everything is stamp collecting.

Sadly, we too often teach evolution and biology as stamp collecting, but integrating evolution and biology would bring concept to otherwise overwhelming complexity.

A theme of this book is that a sophisticated understanding of protein structure and function can be reduced to a simple problem of iterated pattern recognition. I provide many examples.

In the study of ancient evolution, biological concepts leap out and overwhelm. Ancient evolution is a remarkable, beautiful, and coherent story.

The theme of this story encompasses cradles, halos, and wings. TIM barrels (molecular birthday cakes) are something like halos.

Evolution since Coding. http://dx.doi.org/10.1016/B978-0-12-813033-9.00014-7

Chapter 15

Computers in Biology

To appreciate protein structure/function/evolution/dynamics, you need a computer: a PC or Mac will do. You do not need a supercomputer initially.

You are supposed to have loaded VMD, Pymol, YASARA, and/or Chimera on your box (load all 4), and you are supposed to know how to use it (them). Without these essential tools, you cannot appreciate proteins: worse still, you cannot appreciate my book. Without these powerful tools, you cannot appreciate art, because this is the best art. This is also the best science. Perusing a protein structure is like visiting Mars, only more alive, more essential, more human.

Any layman can do this. As I write, I sit as living proof.

You must learn to use the RCSB protein data bank (PDB) to fetch protein sequences (PDB files). You must learn to read the documentation to make sense of the PDB files you download.

Computers can also provide genomic data: cancer databases, genomes, transcription data, epigenetic data, conserved domain data, etc. In a later chapter, I will encourage the use of the online cBioPortal Cancer Genomics Database and the R2 Cancer Database. You will also need to learn to use NCBI online resources. The best way to learn is to run online tutorials.

If you have an interest in biology, your interest must include a computer and the will and wherewithal to use it.

If you have read through previous chapters without learning molecular graphics, return and do it right this time. Eventually, you may thank me.

Evolution since Coding. http://dx.doi.org/10.1016/B978-0-12-813033-9.00015-9

Chapter 16

The Old and New Testaments of Gene Regulation

In one of my scientific studies, I indicated that evolution of life on earth divides naturally into an Old Testament and a New Testament of gene regulation by two double-Ψ–β-barrel-type RNA polymerases. I repeat the analogy in this chapter because I think it is helpful to understand the central importance of RNA synthesis in the evolution of life on earth (Fig. 16.1).

The Old Testament is the chemical generation of self-replicating and coding polymers, the RNA world, the RNA-protein world and the divergence of bacteria and archaea.

The New Testament is the endosymbiotic fusion of a Lokiarchaeota archaea and an α-proteobacterium to form eukaryotes and divergence of eukaryotes (Chapter 31).

In the Old Testament, self-replicating coding ribozymes teamed with ribosomes to advance from the RNA world, dominated by ribozymes, to the RNA-protein world eventually dominated by proteins. Proteins coexisted and were coselected with RNAs and ribozymes. Ribozymes were invaded by proteins, and, in many cases, proteins eventually outcompeted their partner ribozyme to assume the catalytic function. For most reactions in biological systems, proteins (with 20 amino acids and more diverse chemistry) make more capable catalysts than ribozymes (generally with 4 nucleic acids, although many chemical RNA modifications are possible). Many ribozymes of the RNA and RNA-protein worlds, therefore, were replaced by protein enzymes. A key advance was the evolution of a two double-Ψ–β-barrel-type RNA-template-dependent RNA polymerase.

The next momentous event of the Old Testament was the transition from RNA to DNA genomes. Significantly, DNA is a more stable repository of genetic information than RNA. Furthermore, within a cell envelope, intact DNA genomes can be replicated much faster than fragmented RNA genomes that probably cannot be fully encapsulated in cells. On this earth, the great transition from RNA to DNA genomes initially required two double-Ψ–β-barrel-type RNA-template-dependent RNA polymerases and reverse transcriptase, the enzyme that catalyzes conversion of RNA to DNA. Once DNA genomes evolve, a two double-Ψ–β-barrel-type DNA-template-dependent RNA polymerase becomes necessary.

Evolution since Coding. http://dx.doi.org/10.1016/B978-0-12-813033-9.00016-0

55

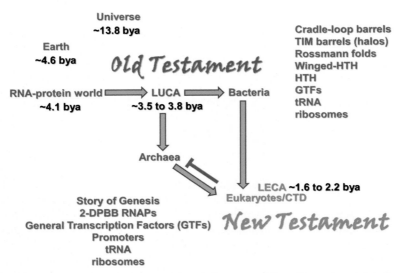

FIGURE 16.1 The Old and New Testaments of gene regulation. The *red symbol* indicates competition between eukaryotes and archaea.

Once DNA genomes became possible, organisms could be fully encapsulated in cells. The first cellular organism with a DNA genome is referred to as LUCA for the last universal common (cellular) ancestor of archaea, bacteria, and eukaryotes. Think of LUCA as the first cellular organism and the first organism with an intact and streamlined DNA genome.

The final chapter of the Old Testament is the great divergence of LUCA to archaea and bacteria, and a simple model is proposed here for the great divergence. Archaea and bacteria have different utilization of two double-Ψ–β-barrel-type DNA-template-dependent RNA polymerases and their general transcription factors. Because, at LUCA, the mechanism of DNA replication is transcription followed by reverse transcription, two double-Ψ–β-barrel-type DNA-template-dependent RNA polymerases and their general transcription factors initially were responsible for both transcription and replication, which are the most central genome maintenance functions. DNA polymerases and separate replication origins are thought to have evolved independently in archaea and bacteria after the great divergence.

The New Testament is the endosymbiotic fusion of a Lokiarchaeota archaea and an α-proteobacterium to form eukaryotes. The colony of invading α-proteobacteria became the mitochondria. The α-proteobacterial genome attacked the Lokiarchaeota genome by unleashing jumping group II intron elements, which gave rise to introns and messenger RNA splicing. In defense against this invasion, the Lokiarchaeota genome generated RNA polymerases I, II, and III, the messenger RNA splicing apparatus, the carboxy terminal domain (CTD) on RNA polymerase II, the cell nucleus, capping, epigenetics, the vast

CTD interactome, complex cell signaling, and many other uniquely eukaryotic innovations. The choices were stark. Eukaryotes either had to innovate or die.

Plants and algae evolved via endosymbiotic fusion of a primitive eukaryote and a photosynthetic cyanobacterium. The invading cyanobacteria left behind the plant chloroplast.

The RNA polymerase II CTD and CTD interactome are central drivers of eukaryote complexity. The Precambrian explosion in animal complexity appears to track with evolution in the CTD interactome to license RNA polymerase II to pause close to the promoter before productively transcribing a gene. Innovation in RNA polymerase II gene regulation tracks with development of (1) complex eukaryotic cell architectures; (2) complex signaling in eukaryotes; (3) regulation of the eukaryotic cell cycle; (4) multicellularity; and (5) animal complexity. Much of the eukaryotic complexity, therefore, arose in defense against the catastrophic invasion of α-proteobacterial group II introns into the Lokiarchaeota genome.

Chapter 17

Evolution as a Cutout Doll Problem III

On the theme of cradles, halos, and wings, we have constructed one barrel (the TIM barrel), and a barrel is almost a halo. Some barrels are shaped like cradles enfolding antiparallel twins.

In keeping with religious themes, let us evolve cradles. Later we can take up wings.

First, some protein origami.

Let us consider monomeric and dimeric RIFT barrels, monomeric double-Ψ–β-barrels, and dimeric swapped-hairpin barrels (Figs. 17.1–17.3). Monomeric double-Ψ–β-barrels are derived from monomeric RIFT barrels, which are derived from dimeric RIFT barrels (now apparently extinct). Let us duplicate a monomeric double-Ψ–β-barrel to generate RNA-template-dependent two double-Ψ–β-barrel-type RNA polymerases, a relic of the RNA-protein world, and then evolve DNA-template-dependent two double-Ψ–β-barrel-type RNA polymerases, which are distributed throughout cellular life. Swapped-hairpin barrels are derived from dimeric RIFT barrels, which appear now to be extinct.

We have considered the inevitability and haunting beauty of $(\beta-\alpha)_n$ repeats (see Chapter 11).

Double-Ψ–β-barrels are another ancient and immortal fold. Double-Ψ–β-barrels resemble cradles holding (antiparallel) twins (Fig. 17.1).

Double-Ψ–β-barrels are among the cradle-loop barrel folds. These include RIFT barrels (Fig. 17.2), double-Ψ–β-barrels (Fig. 17.2), and swapped-hairpin barrels (Fig. 17.3). RIFT stands for occurrence in ribulose isomerase, F1-ATPase and translation factors. A monomeric RIFT barrel is the parent of monomeric double-Ψ–β-barrels. A dimeric RIFT barrel is the parent of dimeric swapped-hairpin barrels, which added an extra β-sheet to each monomer unit to form eight β-sheet barrels from six β-sheet barrels (Fig. 17.3). In Fig. 17.1, a schematic model is shown for the evolution of the cradle-loop barrels. In Fig. 17.2, RIFT barrels and double-Ψ–β-barrels are shown. RIFT barrels are a simple six strand barrel with all antiparallel β-sheets. The RIFT barrel is thought to be generated from a protein dimer of identical β1–β2–α1–β3 units (Figs. 17.1 and 17.2). With time, two identical RNAs may have fused to generate an RIFT barrel as a monomer with a repeated β1–β2–α1–β3–β4–β5–α2–β6 sequence in a single polypeptide chain (Figs. 17.1 and 17.2). Two slightly modified

Evolution since Coding. http://dx.doi.org/10.1016/B978-0-12-813033-9.00017-2

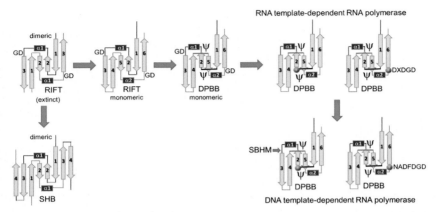

FIGURE 17.1 A schematic model for evolution of cradle-loop barrels. To form the barrels, fold the outside sheets around to interact in the back. β-sheets are *yellow arrows*. α-helices are *purple rectangles*. Mg is shown as a *green sphere*. GD and NADFDGD *boxes* are indicated.

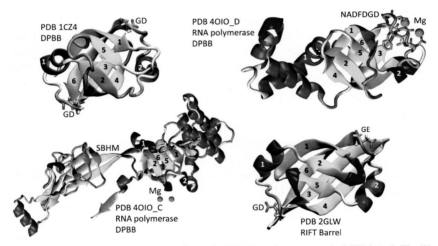

FIGURE 17.2 Monomeric double-Ψ-β-barrels (DPBB) and a monomeric RIFT barrel. The Ψ pattern of crossing chains is indicated for double-Ψ-β-barrels. The double-Ψ-β-barrels of RNA polymerases include insertions (i.e., SBHM for sandwich barrel hybrid motif). The monomeric RIFT barrel is the parent of the monomeric double-Ψ-β-barrel (Fig. 17.1). GD, related GE, and NADFDGD *boxes* are indicated.

double-Ψ-β-barrels found in bacterial *Thermus thermophilus* RNA polymerase (PDB 4OIO) are also shown.

The RIFT barrel is named a cradle-loop barrel because the loops from β1–β2 and from β4–β5 are called the "cradle-loops." The cradle is formed by β1, β6, β3, and β4, and β2 and β5 appear to lie like twin babies in the cradle (Fig. 17.2). The double-Ψ-β-barrel is thought to be derived from

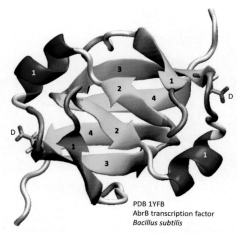

PDB 1YFB
AbrB transcription factor
Bacillus subtilis

FIGURE 17.3 Swapped-hairpin barrels are a cradle-loop barrel posited to be derived from dimeric RIFT barrels, which appear to be extinct. Aspartic acids (D) are derived from GD motifs.

the RIFT barrel by rearrangement of β2 and β5. In the double-Ψ–β-barrel, a complex looping arrangement (a pseudoknot) was generated resulting in the double-Ψ pattern of crossing chains (Figs. 17.1 and 17.2). So, dimeric RIFT barrels became monomeric RIFT barrels probably by ligation of two identical RIFT barrel RNAs. Monomeric double-Ψ–β-barrels were derived from monomeric RIFT barrels by rearrangement of β2 and β5 forming the pseudoknots in the double-Ψ–β-barrel (the crossing Ψ patterns). RIFT and double-Ψ–β-barrels often have GD boxes (glycine-aspartate) just before β3 and β6. The two α-helices (α1 and α2) are evident in the barrel structures. Andrei Lupas and colleagues of the Max Planck Institute are major contributors to structures, analysis, classification, and evolution of the cradle-loop barrel folds.

Dimeric RIFT barrels (now apparently extinct) are thought to be the parents of dimeric swapped-hairpin barrels (Figs. 17.1 and 17.3). Swapped-hairpin barrels added a β-sheet to each monomer unit to form an 8-β-sheet barrel.

Multi-subunit RNA polymerases generate an RNA "transcript" from a DNA template. At the functional heart of multi-subunit RNA polymerases are two double-Ψ–β-barrels; so, there is a duplication of barrel that itself was generated via duplication or gene fusion (Figs. 17.1 and 17.4). A signature motif of multi-subunit RNA polymerases found within the β′ type double-Ψ–β-barrel (NADFDGD; asparagine-alanine-aspartate-phenylalanine-aspartate-glycine-aspartate) includes a characteristic GD box (glycine-aspartate). Despite insertions and modifications, α1 and α2 remain apparent in the two double-Ψ–β-barrels of multi-subunit RNA polymerases (Figs. 17.1 and 17.2). Taking an evolutionary point of view, multi-subunit RNA polymerases should be considered to be two double-Ψ–β-barrel-type RNA polymerases. This part of

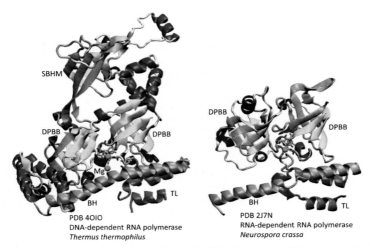

SBHM

DPBB

DPBB   DPBB

Mg

TL

BH

PDB 4OIO
DNA-dependent RNA polymerase
Thermus thermophilus

DPBB

DPBB

BH   TL

PDB 2J7N
RNA-dependent RNA polymerase
Neurospora crassa

FIGURE 17.4 The catalytic cores of two double-Ψ–β-barrel-type DNA-template-dependent and RNA-template-dependent RNA polymerases. SBHM is for sandwich barrel hybrid motif. The bridge helix (BH) is orange. The trigger loop (TL) is red. Mg is green. Positively charged Mg^{2+} is held by acidic groups (i.e., aspartic acid; charge -1).

evolutionary history is clearly written in the genetic code, and it was written between ~3.5 and 4 billion years ago, long before humans put pen to paper. Amino acid sequence and protein secondary structure, therefore, wrote a history of ancient evolution that can still be read. It is remarkable how much of this story remains legible.

In multi-subunit RNA polymerases, on the opposite side of the active site from the two double-Ψ–β-barrels, are the long bridge α-helix and the trigger loop (Fig. 17.4). Multi-subunit RNA polymerases are dispersed throughout archaeal, bacterial, and eukaryotic lineages (so far as I know) without exception. If this existed, organisms lacking RNA polymerases would have a poor prognosis. Universal dispersion of RNA polymerases to the three domains of life anchors multi-subunit RNA polymerases in LUCA (the last universal common ancestor).

But, most likely, RNA polymerases of the two double-Ψ–β-barrel types are more ancient than LUCA, going back into the RNA-protein world. The evidence is the observation that RNA-template-dependent RNA polymerases can also be of the two double-Ψ–β-barrel type (Figs. 17.1 and 17.4). In this case, both barrels are found within a single polypeptide chain. The two barrels have diverged amino acid sequences and therefore separate identities. Reminiscent of DNA-template-dependent multi-subunit RNA polymerases (Fig. 17.4; left panel), these RNA-template-dependent RNA polymerases include a long bridge α-helix and a trigger loop to help frame the active site (Fig. 17.4; right panel).

DNA-template-dependent RNA polymerases polymerize RNA utilizing a DNA template. All archaea, bacteria, and eukaryotes require these to survive, so these important enzymes are distributed to all cellular life on earth.

RNA-template-dependent RNA polymerases polymerize RNA utilizing an RNA template, and these enzymes are now ancient relics of the RNA-protein world. Since DNA genomes took over from RNA genomes, at about the time of LUCA, RNA-template-dependent RNA polymerases of the two double-Ψ–β-barrel type became less essential and have been lost from many lineages. Also, RNA-template-dependent RNA polymerases can be built on multiple genetic plans and need not be of the two double-Ψ–β-barrel type. So far as I know, two double-Ψ–β-barrel-type RNA-template-dependent RNA polymerases have been lost from all archaea, all bacteria, and many eukaryotes. Humans lack a two double-Ψ–β-barrel-type RNA-template-dependent RNA polymerase, but this enzyme is present in some other eukaryotes such as mold *Neurospora crassa* (Fig. 17.4; right panel).

DNA-template-dependent multi-subunit RNA polymerases, however, may retain some memory of their previous functions as RNA-template-dependent enzymes. During hepatitis-δ virus infection (along with hepatitis-B infection; hepatitis-δ virus is a satellite of hepatitis-B virus), human RNA polymerase II partners with hepatitis-δ antigen (encoded by the virus) to replicate and transcribe the hepatitis-δ virus RNA genome. So, in human pathology, human RNA polymerase II can revert to function as an RNA-template-dependent enzyme. This appears to be a genetic memory in the human genome that traces back before LUCA. Also, plant RNA polymerases IV and V, which are two double-Ψ–β-barrel-type RNA polymerases, appear to function in RNA-template-dependent RNA synthesis to generate small regulatory interfering RNAs. Despite the immense gulf in time (~4 billion years), functionally, we of the DNA genome world (including humans) are not as separated from the RNA-protein world as we may think.

Multi-subunit RNA polymerases are among the most beautiful, complex, and dynamic enzymes in the human biosphere. Assembly of the RNA polymerase elongation complex is indicated in Figs. 17.5 and 17.6. In Fig. 17.5, a *Saccharomyces cerevisiae* RNA polymerase II elongation complex (PDB 5C4J) is shown. The right panel indicates the transcription bubble (TDS, NDS, and RNA). Unpaired TDS bases downstream of the active site may indicate preloading of NTP substrates during "transcription" elongation (RNA synthesis). Fig. 17.6 indicates how RNA polymerase is built up around the catalytic core of the two double-Ψ–β-barrels, the bridge helix and the trigger loop.

So, back on the theme of cradles and halos, we have TIM barrels, RIFT barrels, double-Ψ–β-barrels, and swapped-hairpin barrels: all immortal by human standards. All generated as cutout dolls, "LEGO" (Trademark) life and/or protein origami at the earliest times of evolution. All generated as repeats of a simple protein motif. From these core components, some of the most intricate and essential enzymes were evolved.

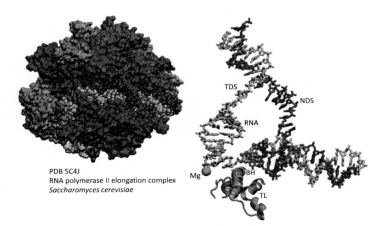

PDB 5C4J
RNA polymerase II elongation complex
Saccharomyces cerevisiae

FIGURE 17.5 A multi-subunit RNA polymerase of the two double-Ψ–β-barrel type. A *Saccharomyces cerevisiae* (yeast) RNA polymerase II transcription elongation complex is shown. Protein subunits are shown in different colors. The right panel shows the transcription bubble in a different view. The template DNA strand (TDS) is lime green. The nontemplate DNA strand (NDS) is magenta. The RNA strand is cyan. The bridge helix (BH) is orange (right image). The trigger loop (TL) is red (right image). Unpaired TDS DNA bases downstream of the active site are colored for chemistry (right image).

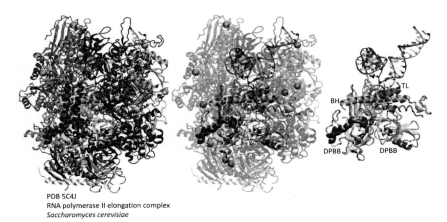

PDB 5C4J
RNA polymerase II elongation complex
Saccharomyces cerevisiae

FIGURE 17.6 The complexity and beauty of two double-Ψ–β-barrel-type RNA polymerases. The leftmost image emphasizes the complex protein subunit structure. The middle image shows the protein subunits rendered transparent to reveal the catalytic core. The rightmost image shows the catalytic core of the double-Ψ–β-barrels (DPBB; colored for secondary structure), bridge helix (BH; orange), and trigger loop (TL; red). Mg is a *green sphere*. Zn are *cyan spheres*. All images are the same view. The TDS is lime green. The NDS is magenta. The RNA is cyan.

Chapter 18

Multi-Subunit RNA Polymerases Book I

Multi-subunit RNA polymerases of the two double-Ψ–β-barrel type are among the most beautiful, complex, and dynamic proteins in the human biosphere. Furthermore, multi-subunit RNA polymerases, their general transcription factors, and promoters form the core of the narrative of evolution of life on earth. In this chapter, I use a bacterial RNA polymerase–initiating complex (PDB 4XLN) interacting with promoter DNA to describe some of the features of these essential enzymes. So, this chapter is an attempt to look under the hood and partly disassemble the perplexing RNA polymerase motor. The RNA polymerase structure I selected is from *Thermus thermophilus*, a bacterial hyperthermophile. A bacterial RNA polymerase was selected because it is slightly simpler than archaeal and eukaryotic RNA polymerases in subunit structure and has fewer zinc (Zn) atoms. Otherwise, because of evolution, features of bacterial RNA polymerase are also features of archaeal and eukaryotic RNA polymerases.

Bacterial RNA polymerase generally utilizes a σ factor as its sole general transcription factor to recognize and open a promoter DNA sequence, and this remarkable event is partly represented in the 4XLN structure with an opened promoter DNA bubble (where the DNA strands separate to expose the DNA template for transcription into RNA). At the active site (where chemistry occurs) RNA polymerase utilizes two Mg atoms (+2 valence each), but only one Mg is seen in the 4XLN structure. Bacterial RNA polymerase includes two Zn ions (+2 valence).

The purpose of this chapter is to give a partial explanation of the functional parts of this mobile RNA polymerization unit, which can be considered to be a molecular motor and also a polymerization factory that synthesizes an RNA copy from a DNA template. Because the enzyme forms a complementary RNA copy synthesized from a DNA template, RNA polymerase makes a "transcript" of DNA and not a "replicate." A replicate would be a complementary DNA copy of a DNA template formed by a DNA polymerase. RNA is chemically distinct from DNA.

Figs. 18.1 and 18.2 show different views of the intricate RNA polymerase structure. Parts of the structure are rendered opaque for particular emphasis. The two double-Ψ–β-barrels and their included loops are drawn in secondary

Evolution since Coding. http://dx.doi.org/10.1016/B978-0-12-813033-9.00018-4

65

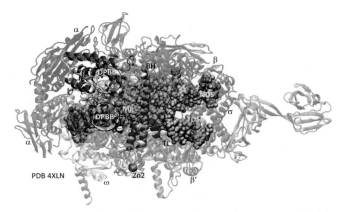

FIGURE 18.1 A *Thermus thermophilus* multi-subunit RNA polymerase initiating complex (subunit structure $\alpha_2\beta\beta'\omega\sigma$) with an open transcription bubble (PDB 4XLN). Parts of the complex image are rendered opaque for emphasis. The two double-Ψ–β-barrels (DPBB) are circled [opaque and colored for secondary structure (i.e., β-sheets are yellow, α-helices are purple)]. The trigger loop is red (opaque). The bridge helix is orange (opaque). Mg (magnesium) at the active site is a magenta sphere. Zincs (Zn1 and Zn2) are cyan spheres. The template DNA strand (TDS) is green. The nontemplate DNA strand (NDS) is gray. Subunits are in different colors and labeled (transparent).

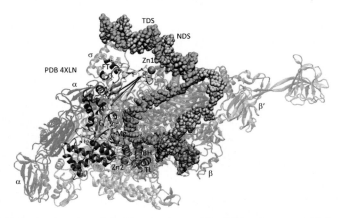

FIGURE 18.2 Another view of the *Thermus thermophilus* RNA polymerase initiation complex (PDB 4XLN).

structure representation (opaque). The bridge helix (BH) is orange and opaque. The trigger loop (TL) is red and opaque. Looking at these structures is a bittersweet experience for me, because I thought I would die long before such beautiful, informative, and intricate images became available. Some of the features and potential functions of the RNA polymerase structure are described below.

DOUBLE-Ψ–β-BARRELS (THE ACTIVE SITE)

Based on evolution, dynamics, and location, a simple model (2017) for the function of the two double-Ψ–β-barrels is proposed. The double-Ψ–β-barrels are posited to form the core of the stepwise, thermal ratchet to move (translocate) nucleic acids (DNA and RNA) through the RNA polymerase structure. RNA polymerase is a molecular motor that must make accurate single base steps along its DNA template to synthesize RNA. RNA polymerase, therefore, can be considered to be a mobile factory to synthesize RNA from a DNA template. Restrained, single-step motor action is also supported by the bridge helix and trigger loop, which also frame the RNA polymerase active site. The location of the active site, where RNA polymerase does chemistry, is identified by the Mg (magnesium) atom (+2 valence). The two double-Ψ–β-barrels, therefore, can be thought of as both a motor to translocate RNA/DNA and as a restraint to prevent RNA polymerase from accidentally slipping to take multiple steps rather than single steps. Each baby step by RNA polymerase is a single base unit (~3.4 Å) and must be limited to a single step.

The two Mg atoms (+2 charge each) are held at the RNA polymerase active site (where chemistry occurs) by acidic (−1 charge each) amino acid residues [aspartic acid (D) and glutamic acid (E)] (Fig. 18.3). Because, in the 4XLN structure, RNA polymerase was not caught in the act of trying to form a phosphodiester bond (the polymerization reaction), only one Mg is observed. The signature sequence of multi-subunit RNA polymerases is NADFDGD (asparagine-alanine-aspartate-phenylalanine-aspartate-glycine-aspartate), which includes a GD box (glycine-aspartate) typical of RIFT barrels and double-Ψ–β-barrels (Chapter 17). The 737-NADFDGD-743 sequence is found in the β′ subunit of RNA polymerase.

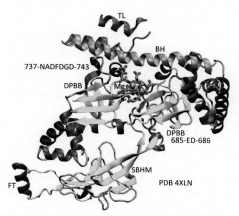

FIGURE 18.3 The RNA polymerase active site is shown between the two double-Ψ–β-barrels (DPBB), the bridge helix (BH), and the trigger loop (TL). The sandwich barrel hybrid motif (SBHM) is inserted into a DPBB loop. FT is the "flap tip" helix, an extension of the SBHM. The active site Mg atoms are held by acidic residues (D and E; charge −1).

The three Ds (negatively charged aspartic acid or aspartate) hold the Mg atom (+2 charge) in the structure. The 685-ED-686 sequence from the β subunit helps to hold the second Mg atom (not present in the 4XLN structure).

Because both β and β′ subunits of RNA polymerase include a double-Ψ–β-barrel, RNA polymerase was evolved via a genetic sequence duplication that took place in the RNA-protein world. The active site of RNA polymerase is formed by the two double-Ψ–β-barrels, the bridge helix, and the trigger loop.

THE BRIDGE HELIX

The bridge α-helix is a long helix (~38 amino acids) that seems to form a "bridge" between the β and β′ subunits. The bridge α-helix is a major feature of the β′ type subunit of multi-subunit RNA polymerases. Thermal motions of the bridge helix are thought to aid stepwise translocation of RNA/DNA through the RNA polymerase structure.

THE TRIGGER LOOP

The trigger loop is another prominent feature of the β′ subunit of RNA polymerase. The trigger loop is highly dynamic with "open" [translocating and nucleoside triphosphate (NTP) loading] and "closed" (catalytic) conformations. In PDB 4XLN, the conformation of the trigger loop is open, and some of the central trigger loop is disordered and therefore not visible in the X-ray crystal structure. An open trigger loop structure does not support the chemical step in RNA synthesis. Rather, the open trigger loop has been associated with the loading of NTP substrates into the active site for RNA synthesis and with the translocation (forward sliding) step. A closed trigger loop structure (not shown) is the catalytic form of RNA polymerase.

Because the trigger loop is open in 4XLN, this may contribute to the somewhat disordered state of the β subunit double-Ψ–β-barrel, in which β2 and β5 of the six β-sheet barrel were not scored by the Visual Molecular Dynamics program as β-sheets. In a catalytic structure with a closed trigger loop (i.e., PDB 2O5J) the β subunit double-Ψ–β-barrel is more ordered, and all 6-β-sheets are apparent in the barrel (i.e., see Chapter 17).

σ FACTOR BINDING TO PROMOTER DNA

Fig. 18.4 shows the general transcription factor σ binding to the RNA polymerase promoter and interacting with the RNA polymerase catalytic core. The rest of the 4XLN structure is hidden in the image. σ is modeled as a four HTH factor. Helix-turn-helix (HTH) domains are more nearly "helix-turn-helix-turn-helix" domains (H1-T1-H2-T2-H3) (H for helix; T for turn). H3 is referred to as the "recognition" helix. The N-terminal end of H3 generally binds in the DNA

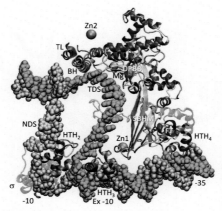

FIGURE 18.4 Interactions of the RNA polymerase catalytic core, the promoter DNA and the σ general transcription factor. The σ factor is gold transparent. Three helix-turn-helix (HTH) domains that interact with the promoter DNA sequence are rendered opaque for emphasis. The flap tip (FT) helix binds to σ HTH_4 that binds the −35 region of the promoter. σ HTH_3 binds where the transcription DNA bubble opens (Ex −10 for Extended −10 region). σ HTH_2 opens the promoter at the −10 sequence. σ HTH_1 is vestigial and no longer makes strong promoter contacts.

major groove on double-stranded DNA and typically makes most sequence-specific contacts to DNA. H1 and H2 tend to form a structural brace for H3. The flap tip (FT) helix of the SBHM interacts with σ HTH_4, so this interaction links the β subunit double-Ψ–β-barrel to the σ factor and −35 region of the promoter (the "anchor" DNA sequence). σ HTH_3 interacts where the DNA bubble opens. σ HTH_2 opens the promoter −10 region by flipping out bases on the nontemplate DNA strand. I thought I would die long before I was able to imagine a spare and compact model for σ factor functions in initiation, but here I present a simple model that integrates evolution, structure, and dynamics.

ZN1 AND ZN2

Archaeal and eukaryotic RNA polymerases include many Zn domains, but bacterial RNA polymerases have two Zn atoms, here labeled Zn1 and Zn2, both in the β′ subunit. Zn1 (Fig. 18.5) is near the N-terminus of the β′ chain. This is a fairly typical Zn-binding domain. The Zn atom (+2 valence) interacts with four cysteine residues (−1 charge each). The Zn1 domain interacts with the exiting RNA chain during RNA chain elongation. The Zn2 domain is a somewhat less typical Zn-binding domain with a role in coupling the interactions and dynamics of the bridge helix and the trigger loop. Zn2 (+2 valence) is also held by four cysteine atoms (Cys 1112, Cys 1194, Cys 1201, and Cys 1204) (−1 valence each). One strange feature of the Zn2 domain is the large amino acid spacing between Cys 1112 and Cys 1194 (80 amino acids). The expected spacing is two amino acids (CXXC; C = Cys, X = any amino acid) (Fig. 18.6).

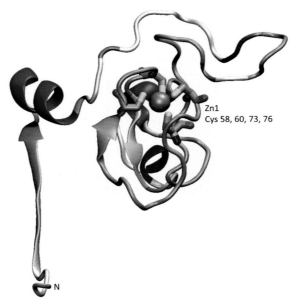

FIGURE 18.5 The Zn1 (*silver sphere*) domain is close to the N-terminus of the β′ subunit chain. Zn1 is held by Cys 58, Cys 60, Cys 73, and Cys 76. The yellow atoms of cysteine are sulfurs that bind Zn. Zn1 is part of the RNA exit pore.

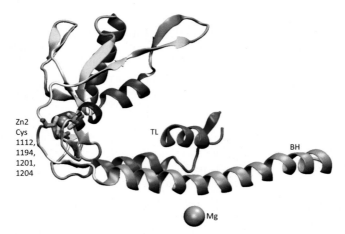

FIGURE 18.6 Zn2 connects the bridge helix and the trigger loop, which appear to work together in catalysis (chemistry) and translocation (RNA/DNA movement). Cys 1112, Cys 1194, Cys 1201, and Cys 1204 bind Zn2 (*silver sphere*). The yellow atoms of cysteine residues are sulfurs (charge −1 to bind Zn^{2+}). Mg (active site) is a magenta sphere.

In the β' chain, I am neglecting (among other things) the β-hairpin (β' 91–100 and 513–519), the two coiled-coils (β' CC1 539–607 and CC2 958–1014), and the AT-hooks (β' 583–602) are found in all multi-subunit RNA polymerases, so these additional features are also present in RNA polymerase at the last universal common ancestor (LUCA). In the β chain, I neglect the fork loop (β 400–456) also present since LUCA. Multi-subunit RNA polymerases are cobbled together similar to a kind of Frankenstein's monster of connected parts. These parts, apparently, were linked via evolution and by the rules of genetics rather than by a supreme being. Frankenstein's monster, by contrast, could be considered to be "intelligently designed."

DO NOT TRY THIS AT HOME

If you wish to confirm the structures in this chapter using a laptop computer and the program Visual Molecular Dynamics, you may have to do a few things. I did not do anything very fancy, but I did edit the PDB 4XLN text file to make it as small as possible to help my little computer cope. PDB files can be edited as simple text files, but you may have to gain some insight into the PDB text file structure (i.e., chain designations) to make useful edits. There are two RNA polymerase molecules in the original PDB file and two sets of nucleic acid chains. To reduce the size of the file, I kept one molecule and deleted the other using Pymol. Otherwise, the PDB file is too big to be comfortably manipulated on a small computer using VMD. PDB 2O5J (with a closed trigger loop and loaded NTP substrate) is another interesting RNA polymerase structure to examine, but this file may also be too large for a small computer and may need editing. Pymol is very convenient for reducing the size of large PDB files.

The only way to break down this complex information is using molecular graphics. Without these tools, it is impossible to visualize RNA polymerase structure and function. Two dimensional images can only partially represent the beauty and complexity.

Chapter 19

Multi-Subunit RNA Polymerases Book II

Chapter 18 was a look under the hood of the two double-Ψ–β-barrel-type RNA polymerases to peek at the motor and RNA production factory. This chapter gives broader insight into the family, functions, peripherals, evolution, radiation, and diversity of two double-Ψ–β-barrel-type RNA polymerases.

Carl Woese demonstrated that ribosomes form the core of evolution. More about this later.

However, as I heard Max Perutz so wisely explain, "Of course hemoglobin has the same structure in the crystal and in solution. I wouldn't have spent my life working on it if it didn't."

Let us leave ribosomes for the moment and concentrate on something really, really important, but not hemoglobin. Let us concentrate on multi-subunit RNA polymerases of the two double-Ψ–β-barrel type.

The major sign posts in evolution are the transition from the RNA-protein world to the DNA genome and cellular world (LUCA), divergence from LUCA to archaea and bacteria, and then unholy fusion of archaea and bacteria to form eukaryotes (the last eukaryotic common ancestor). By comparison, human evolution is a detail (an endnote).

In addition, the ribosome is not changed much since the RNA-protein world, making the ribosome a somewhat less informative marker for the major sign posts in evolution.

By contrast, the story of RNA polymerases of the two double-Ψ–β-barrel type is the story of life on earth: this is the core story of genesis. RNA polymerases tell the story of the RNA world progressing to the RNA-protein world. RNA polymerase tells the story of the transition from RNA to DNA genomes at LUCA. RNA polymerase tells the story of the divergence of archaea and bacteria. RNA polymerases tell the violent and passionate story of the birth of eukaryotes. RNA polymerases tell the story of eukaryote complexity. What more would you like to know? What more is there to biology?

This chapter gives somewhat of an overview from the point of view of RNA polymerases. Subsequent chapters give more details.

In the beginning of coding, it is thought that RNAs may have been self-replicating. The precise mechanism cannot now be fully known, although attempts to generate RNA enzymes that will accomplish primed RNA-template-dependent

Evolution since Coding. http://dx.doi.org/10.1016/B978-0-12-813033-9.00019-6

replication and transcription are underway. Whether or not ancient evolution can be recreated, however, it is very likely that such attempts will succeed, showing that such an evolutionary scheme is feasible.

At the dawn of the RNA-protein world, ancient ribosomes evolved to synthesize proteins, directed by RNA templates.

Proteins began to invade ribozymes. Initially, proteins were selected as protein cofactors of RNA enzymes, but, with time and evolution, proteins began to assume many ribozyme functions.

Although a failed attempt, the process of proteins invading a ribozyme is apparent in ribosome structures (Fig. 19.1). The leftmost image is of a human ribosome with proteins. In the cente rimage, proteins are removed. The rightmost image shows tRNAs bound to mRNA and indicates the peptidyl transferase center where protein is synthesized. The capacity of ribosomes to support protein synthesis remains embedded in ribosomal RNA, so the ribosome can be considered to be a large ribozyme: a catalytic RNA. Although studded with protein cofactors, catalytic potential of the ribosome remains specific to the peptidyl transferase center, which is composed of rRNA, as it was at the birth of the RNA-protein world. An interesting hypothesis for why proteins have not taken over ribosomal RNA ribozyme functions in protein synthesis, according to Aravind (NCBI), is that RNA catalysts are more talented than protein catalysts at polypeptide synthesis, and proteins, therefore, never managed to supplant the ribosome ribozyme functions. Strangely, proteins appear to be not as talented as one of the largest ribozymes (the ribosome) at synthesizing peptide oligomers. An alternate explanation might be that molecular clustering of amino acylated tRNAs in the peptidyl transferase center might be sufficient to support peptide bond synthesis on the ribosome and that there was no strong evolutionary pressure to improve this ribosome function. Mutagenic analyses indicate that there is little specific catalytic function to the peptidyl transferase

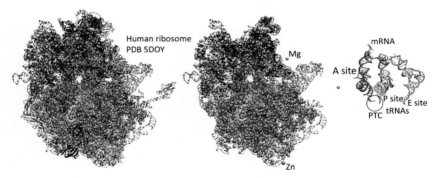

FIGURE 19.1 The ribosome as a battleground between protein cofactors and ribozyme functions. Protein cofactors are shown only in the left image. Mg is pink (*spheres*). Zn is cyan (*spheres*). The rightmost panel shows the tRNAs in the A site (red), P site (orange), and E (exit) site (green) interacting with mRNA (cyan). PTC indicates the peptidyl transferase center where protein is synthesized.

center of the ribosome. For whatever reason, however, proteins failed to replace peptidyl transferase functions, and ribosomes remain large ribozymes studded with many protein cofactors.

By contrast, proteins make very adequate RNA polymerases, and, in this case, protein invasion of a self-replicating ribozyme was successful. Two double-Ψ–β-barrel type RNA-template-dependent RNA polymerases were born. Probably, these enzymes looked very much like PDB 2J7O interfering RNA polymerase from *Neurospora crassa* (Fig. 17.4) with two double-Ψ–β-barrels holding two catalytic Mg^{2+} ions and also a bridge helix and a trigger loop.

LUCA is discussed here as the first cellular organism with a DNA genome. I understand that this is an oversimplification.

To go from an RNA genome world to a DNA genome world one requires (1) RNA-template-dependent RNA synthesis (RNA polymerase) and (2) RNA-template-dependent DNA synthesis (reverse transcriptase). So, to approach cellular life on this earth, you require a two double-Ψ–β-barrel-type RNA-template-dependent RNA polymerase. Then you need to replace this function with a two double-Ψ–β-barrel-type DNA-template-dependent RNA polymerase. To make DNA from RNA, you require a reverse transcriptase, as in retroviral replication today.

The evolutionary driving force from RNA to DNA genomes is clear. DNA confers higher genomic stability (Fig. 19.2). DNA is chemically more stable than RNA. On its sugar ring, DNA lacks a 2′-OH, present in RNA and highly reactive with the 3′-O linkages that form the nucleic acid chain. Hydrolysis of RNA, therefore, is much easier than for DNA, particularly in alkaline (basic; >pH = 7) solution. The double-strand character of DNA shields bases

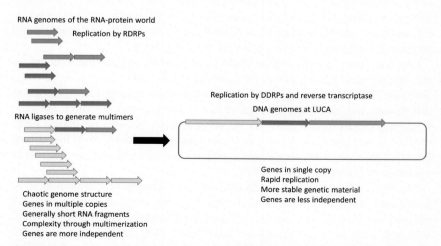

FIGURE 19.2 The evolution of DNA genomes from RNA genomes. *DDRPs*, DNA-template-dependent RNA polymerases; *RDRPs*, RNA template-dependent RNA polymerases.

from reaction and builds redundancy in coding. DNA is a much more stable genetic material than RNA.

The first large DNA genomes brought another advantage but at a price. The advantage of large DNA genomes is that they can be replicated efficiently and rapidly, decreasing generation times to generate many progeny. The potential downside of the DNA genome is that genes become much more interdependent. By contrast, RNA genomes are small and volatile with chaotic and error-prone replication. So, the RNA-protein world may have comprised multiple small RNA units, with variable gene copy numbers and much greater gene independence. For this reason, the RNA-protein world may have been a time of accelerated genetic innovation compared to the DNA genome world. Because, in the early DNA genome world, rapid replication becomes a big advantage, individual gene copy numbers are low (usually single copy), genes are highly interdependent and replication rates are fast. Also, an organism with an intact and comprehensive genome can be encapsulated in a cell within a membrane. It is difficult to imagine how a fragmented RNA genome could be fully compartmentalized.

A major issue in rooting the tree of life at LUCA and evolving divergence to archaea and bacteria has been that archaea and bacteria have distinct genetic plans for DNA genome replication: notably archaea and bacteria utilize different (nonhomologous; not related) DNA polymerases.

There might be many explanations, but this is the one I favor (2017). LUCA used the replication mechanism of transcription followed by reverse transcription, required to advance from an RNA genome world to a DNA genome world. After divergence of archaea and bacteria, archaea evolved a more efficient and processive DNA polymerase mechanism, and bacteria, separately, did the same, explaining separate and nonhomologous systems for DNA replication in archaea and bacteria. The evolutionary decision to diverge archaea from bacteria was made on the basis of RNA synthesis, a more ancient and more central process in evolution of life on earth than DNA synthesis.

Archaea have a more ornate multi-subunit RNA polymerase than bacteria. The archaeal RNA polymerase more closely resembles that of eukaryotes. Bacterial RNA polymerase is more streamlined. There must be reasons for these differences, but, currently, it is not clear to me the extent to which bacterial RNA polymerase lost complexity or archaeal RNA polymerase gained complexity. In any event, divergence of multi-subunit two double-Ψ–β-barrel-type RNA polymerases track divergence of archaea and bacteria, a major signpost in evolution of life on earth (Fig. 19.3).

In Fig. 19.3, a map and crude time line of evolution of two double-Ψ–β-barrel-type RNA polymerases is shown. An RNA-template-dependent RNA polymerase evolved during the RNA-protein world. Such RNA polymerases still exist in some eukaryotes today (i.e., PDB 2J7N). RNA-template-dependent RNA polymerases lack the sandwich barrel hybrid motif (SBHM) found in DNA-template-dependent RNA polymerases. The sandwich barrel

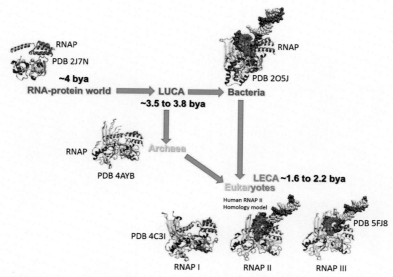

FIGURE 19.3 Evolution of RNA polymerases. Three RNA polymerases in eukaryotes. One in bacteria. One in archaea. Only the catalytic core of the RNA polymerase (RNAP) is shown.

hybrid motif, therefore, is considered an adaptation for utilization of a DNA template. At LUCA, there was a switch to a DNA-template-dependent RNA polymerase. Both archaea and bacteria have a single two double-Ψ–β-barrel-type DNA-template-dependent RNA polymerase used to transcribe all archaeal and bacterial genes. Eukaryotes have at least three two double-Ψ–β-barrel-type DNA-template-dependent RNA polymerases. In Fig. 19.3, only the catalytic core of two double-Ψ–β-barrel-type RNA polymerases is shown. The catalytic core of RNA polymerases is here primarily considered to be the two double-Ψ–β-barrels, the bridge helix, and the trigger loop.

Eukaryotes are an unholy fusion of archaea and bacteria to be explained in more detail and with more drama below. Generation of eukaryotes resulted in an explosion of two double-Ψ–β-barrel-type RNA polymerases. As noted above, eukaryotes have at least three two double-Ψ–β-barrel-type RNA polymerases, RNA polymerases I, II, and III (Figs. 19.3 and 19.4). RNA polymerase I synthesizes ribosomal RNA. RNA polymerase II synthesizes messenger RNA and some small, regulatory RNAs. RNA polymerase III synthesizes transfer RNAs, 5S RNA, and some small regulatory RNAs. This allows messenger RNA synthesis, which is generally most important in terms of gene expression control, to be uncoupled from ribosomal RNA, transfer RNA, and 5S RNA synthesis. Plants have RNA polymerases IV and V, which are involved in interfering RNA synthesis. In Fig. 19.4, the single bacterial and archaeal RNA polymerases are shown for comparison to eukaryotic RNA polymerases I, II, and III. RNA

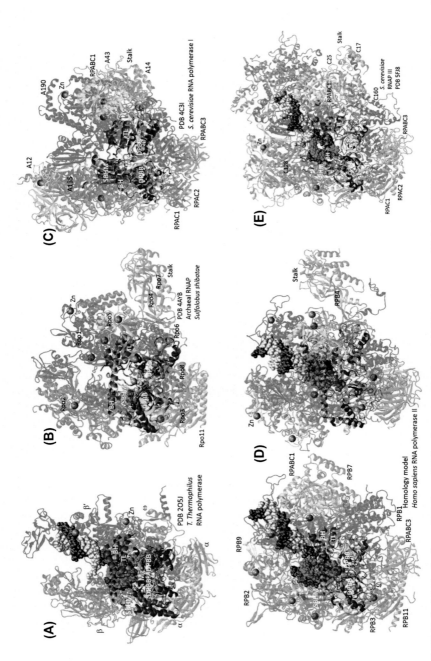

FIGURE 19.4 Multi-subunit two-DPBB type RNA polymerases. (A) Bacterial RNA polymerase. (B) Archaeal RNA polymerase. (C) Yeast RNA polymerase I. (D) Human RNA polymerase II (2 views). (E) Yeast RNA polymerase III. Colors are selected to emphasize homology of features and subunits. The CTD on RNA polymerase II is largely unstructured and is not visualized in X-ray crystal structures.

polymerase subunits are colored to indicate homology (genetic relatedness) among subunits in different organisms and RNA polymerases.

Messenger RNAs (mRNAs) are those that are read on the ribosome to specify protein synthesis (Fig. 19.1), so, in a complex eukaryote, specifying mRNA synthesis in any particular cell type is key to conferring the identity of that particular cell. In order to further differentiate RNA polymerase II transcription and mRNA synthesis, RNA polymerase II became decorated with the strange carboxy terminal domain (Fig. 19.5).

Because you understand $(\beta-\alpha)_n$ repeat proteins (Chapter 11), you understand genetic multimerization. In the case of the carboxy terminal domain (CTD) of RNA polymerase II, think about the strange CTD extension as a scaffold on RNA polymerase II allowing evolutionary innovation coupled specifically to mRNA synthesis (Fig. 19.5). Similar to $\beta-\alpha$ repeat proteins, the CTD is a protein repeat sequence [1-YSPTSPS-7: tyrosine (Y), serine (S), proline (P), threonine (T), serine (S), proline (P), serine (S)]. In yeast, there are 26 or 27 repeats, and many of them are perfect YSPTSPS repeats. In humans, there are 52 repeats, more than half of them are perfect YSPTSPS repeats.

Evolutionary biologist John Stiller (East Carolina University) has explained that the CTD probably initially evolved to cope with mRNA splicing. I believe Stiller is correct in this. This is a thrilling story that will be picked up again.

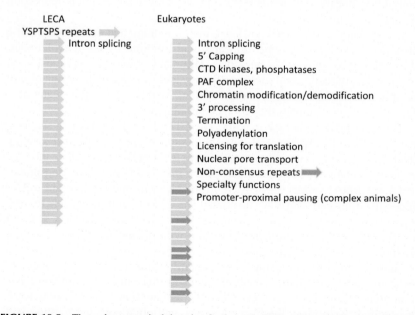

FIGURE 19.5   The carboxy terminal domain of eukaryotic RNA polymerase II as a scaffold for evolutionary innovation. Initially, the YSPTSPS repeats supported intron splicing. Since the last eukaryotic common ancestor (LECA), many other systems have coevolved with the CTD to support transcriptional functions.

With time, many mRNA-related activities evolved to hitchhike on the CTD. Because the CTD is a topic for a lifetime of study, a complete description of the vast and mysterious CTD interactome is not provided here. The CTD is at the heart of RNA polymerase II recruitment to a promoter, diverse promoter functions, mRNA capping, mRNA splicing and polyadenylation, cell-signaling functions, chromatin modification systems, interactions between active genes and the nuclear pore, mRNA export from the cell nucleus, and licensing of mRNA for translation on the ribosome. In some ways the CTD acts as a kind of global positioning system informing RNA polymerase II whether it is initiating, pausing, elongating, terminating, or recycling. From another view, the CTD is a scaffold for many factors that must interact with RNA polymerase II transiently and then release. The CTD is a miracle of, and a poster for, evolutionary "design" of cycles. The CTD is a repeating $(YSPTSPS)_n$ sequence scaffold attached to RNA polymerase II that is coevolved with a myriad of recruited interacting factors that support the transcription cycle (initiation, promoter escape, elongation, termination, recycling).

Looking at the RNA polymerase II CTD from the point of view of intelligent design versus evolution, the CTD and its vast interactome appear to be evolved. As with $(\beta-\alpha)_n$ repeats and double-Ψ-β-barrels, the CTD is a protein cutout doll pattern [i.e., human $(\sim YSPTSPS)_{52}$]. The CTD interactome somewhat resembles a Rube Goldberg device of seemingly endless add-ons and compensations. Everything appears to be a functional add-on of reversible binding, repression, and antirepression functions.

It is perhaps useful to acknowledge that in biology and evolution, smart is often stupid. That is: "designs" appear stupid followed by harsh evolutionary editing that gives purpose in retrospect. Ultimately, biology cares most about successful reproduction and survival. Biology is less fussed with elegant design, and it does not do planning or design upfront. Without purposeful planning at the onset, evolution applies the rationale through killing off failures.

The CTD might be a case in point. The CTD is a dumb repeating sequence attached to RNA polymerase II initially to support cotranscriptional splicing, which, in fact, can occur without tethering to RNA polymerase. Once the CTD becomes available, it becomes a part of the milieu. Many hitchhiking protein (and some RNA) factors then evolve to interact with the CTD to aid mRNA processing, transport, and other functions. The CTD, therefore, appears to make sense only in the light of evolution: the CTD is an evolved repeat scaffold, refined and developed through coevolution with organisms, gene expression, and hitchhiking exchange factors.

Humans think they are smart. With regard to human and animal memory, however, from the biological point of view, too smart may be stupid, because, to an animal, sometimes forgetting may be as important to survival as learning. Animals have many inhibitory neurons, and often repeated training with intermediate resting is required to fix memories into long-term storage. So, the threshold for retaining information is often quite high in animals, and less important information is given low priority for retention. Survival, therefore, can be supported by forgetting rather than learning. From an evolutionary view, for an animal, being too smart may not be most survivable, and, in a nuclear age with growing international fascism, humans may be too smart for their own good.

Most eukaryotes are more complex than archaea and bacteria.

Why is this so and what are the consequences?

Multi-subunit two double-Ψ–β-barrel-type RNA polymerases and their general transcription factors are more complicated in eukaryotes than in archaea and bacteria, and eukaryotes have multiple RNA polymerases (minimally, RNA polymerases I, II, and III). Eukaryotes have the CTD on RNA polymerase II, which is absent in archaeal and bacterial RNA polymerases. In eukaryotes, the CTD appears to license more complex mRNA regulation involving coevolution of the CTD and its interactome. The CTD is powerfully coevolved with cell-signaling systems, which are important for cell communications and organism complexity. Multiple RNA polymerases and pinning the CTD on RNA polymerase II were among the events that allowed eukaryotes to become more complex than archaea and bacteria. More complex systems for gene regulation, particularly for mRNA synthesis, were necessary to generate organisms of increasing complexity.

Chapter 20

The Chemical Synthesis of Life

Trained as a molecular biologist/biochemist, I tend to think in terms of molecular coding, so, I may be the wrong person to tell this ancient story.

I believe that the major concepts may reduce to the following:

Generate chemical polymers as a basis for future coding.

Generate or harness battery power (oxidation–reduction power) as a basis for future metabolism, as in the citric acid cycle and the glyoxylate cycle. Prelife and chemical polymerization reactions require energy transduction.

Generate chemical energy similar to ATP and GTP hydrolysis.

From this, it should be possible to evolve life on earth: RNA coding, ribozymes, protein, and eventually mRNA-coded protein synthesis by a ribosome.

Nick Lane tells many fascinating stories of ancient energy transduction in biological systems in more detail in his book *The Vital Question*.

In a recent paper (2017), it was reasonably proposed that the last universal common ancestor (LUCA) (the first cellular life) utilized the Wood–Ljungdahl pathway to harness evolving hydrogen gas to fix CO_2 and form acetic acid ($4H_2 + 2CO_2 \rightarrow CH_3COOH + 2H_2O$). In such a reaction, redox power from H_2 is utilized to reduce CO_2 to provide redox power and a reduced carbon source to drive metabolism. According to this view, LUCA was highly dependent on a particular environment, such as a deep sea hydrothermal vent evolving hydrogen gas and carbon dioxide, to support its initial foray into cellular life.

From a molecular biologist's point of view, it is difficult to look back to the chemical synthesis of life and the dawn of biological energy transduction. It seems more natural to look back to the RNA or the RNA-protein world and the birth of ribozymes and coding. From the RNA or RNA-protein world, ancient evolution becomes much more familiar.

I talked briefly with Aravind (NCBI) about panspermia, the dispersion of life throughout the universe. But I do not find that panspermia is necessary to explain evolution of life on earth. The universe is ~13.8 billion years old. Earth is ~4.6 billion years old. So, evolving life elsewhere in the universe and distributing it seems possible but not really necessary or that helpful. Probably ~4.1 billion years is sufficient.

Also, living systems appear to explode from the RNA-protein world to the DNA genome world and LUCA. Remarkably, this rapid transition appears to have taken only a few hundred million years (i.e., ~4.1 to ~3.8 billion years ago and the window appears to continue to tighten). By contrast to the DNA genome

Evolution since Coding. http://dx.doi.org/10.1016/B978-0-12-813033-9.00020-2

world, the RNA-protein world appears to be a crucible of remarkable evolutionary innovation. After the advent of rapidly replicating and streamlined DNA genomes, evolution appears to become much slower and more conservative. The window for the RNA-protein world appears to narrow as more is learned about the transition to LUCA. I have argued that the transition to more stable and streamlined DNA genomes makes genes more codependent, slowing some evolutionary processes.

I have read stories about universe "inflation." It is my understanding that microwave telescope experiments that seemed initially to strongly support inflation theory have been criticized as a probable artifact, but this does not contradict inflation theory, which still may be correct. Inflation maintains the universe at a fairly constant temperature, which might have supported panspermia when the universe was at a higher consistent temperature. Now the universe appears to be too cold to support ongoing dispersion of life throughout a rapidly expanding universe.

I find these concepts fascinating but beyond my ken.

Chapter 21

The RNA World

In the chapter on chemical synthesis, we generated dumb polymers, energy generation, and primitive metabolism: battery power (redox chemistry) and chemical energy (analogous to ATP and GTP).

Probably, some kind of sequestration and compartmentalization may also have been required, particularly for energy generation, which requires maintenance of a proton gradient across a membrane.

The distant RNA world can largely be built on such scaffolds.

Primitive polymers must be swapped to become RNA polymers.

RNA must then fold into ribozymes.

Ribozymes still exist today: the ribosome peptidyl transferase center, RNase P, the spliceosome, the hammerhead, hepatitis-δ virus, self-splicing introns, many others.

Functional RNA scaffolds are still apparent: tRNAs, 5S RNA, ribosomal RNA, others.

Whether a world dominated by ribozymes, preribosomes, and primitive RNA coding is easy to imagine or not, one can still see significant parts of it.

We have hypothesized a self-replicating RNA ribozyme with gene coding potential. This idea comes from Aravind (NCBI) and others, not from me. This ribozyme's RNA polymerase activity is to be later assumed by a protein two double-Ψ–β-barrel-type RNA-template-dependent RNA polymerase.

Because we can see so many relics of the RNA world, it is not so difficult to imagine such a time. It was a time dominated by RNA coding, diverse ribozymes, probably RNA chemical modifications to broaden ribozyme chemical activities, and primitive metabolism, as described for the chemical synthesis world (Chapter 20).

Transfer RNAs have many chemical modifications that are fundamental to function and specificity. One therefore can imagine a time when ribozymes possessed much more diverse chemistry than they are known to possess today. Transitions to protein catalysts and the DNA genome world may have enforced greater genetic conformity on RNA than was in place in a primitive RNA-dominated world.

Once the primitive ribosome or proto-ribosome is generated, we begin to move to the more familiar, but still distant, RNA-protein world.

Evolution since Coding. http://dx.doi.org/10.1016/B978-0-12-813033-9.00021-4

Chapter 22

Ribosomes

Ribosomes are often considered to be ribozymes (RNA enzymes) that synthesize proteins, which are polymers of amino acids. RNA is a single-stranded polymer of A, G, C, and U bases. Ribosomes are mostly made up of ribosomal RNA (rRNA) with multiple protein subunits (Fig. 19.1).

At its core, life is largely made up of template-directed polymers. Initially, this was RNA with little or no protein. RNA formed the catalysts (ribozymes, RNA enzymes) and also was responsible for coding (genetic memory).

The ribosome is a ribozyme that can read an RNA template (a messenger RNA) to generate a protein amino acid polymer. As such, the ribosome bridges the RNA coding world to the RNA-protein world. Ribosomes or some proto-ribosomes must be about 4 billion years old on earth.

Transfer RNAs (tRNAs) are necessary to bring amino acids one by one to the ribosome for joining to a growing polypeptide chain (Fig. 19.1). Cloverleaf tRNAs may have evolved to replace a more primitive proto-tRNA adapter molecule. Francis Crick hypothesized the existence of adapter molecules for translation (protein synthesis), and he and others were delighted and surprised to discover the complexity of tRNAs. They had expected a simpler adapter.

Once proteins become available, they coevolve with RNAs to do many jobs. As catalysts, proteins are often more talented than ribozymes. Proteins comprised 20 amino acids as opposed to 4 bases (A, G, C, U) of RNA. Therefore proteins are capable of folding to generate more chemical capacity and complexity.

As Aravind (NCBI) has pointed out, in 4 billion years of evolution, proteins have failed to replace ribosomes as amino acid polymerases. Stubbornly, the ribosome persists as a primitive ribozyme, at least for its peptidyl transferase function. But proteins have tried with countless failures to take over the amino acid polymerase function. Ribosomes are studded with these failed protein cofactors (Fig. 19.1). By contrast, RNA polymerases of the two double-Ψ–β-barrel type were much more successful in replacing their ribozyme. Because of the chemical complexity and complex folds of proteins, most catalysts in modern biology are protein catalysts rather than RNA catalysts. Only a few biological catalysts persist as ribozymes. The ribosome may be the most central, essential, and enduring ribozyme, but some argue that the ribosome is hardly a ribozyme at all.

Evolution since Coding. http://dx.doi.org/10.1016/B978-0-12-813033-9.00022-6

So, at the risk of contradicting myself, the ribosome peptidyl transferase center does not appear to be a very good or efficient ribozyme. Current thinking (2017) is that the peptidyl transferase center is mostly a clustering environment for the amino acylated 3'-ends of transfer RNAs and that bringing two amino acylated tRNAs close enough together may be sufficient to support peptide bond formation.

Because of its higher stability, DNA has come to replace RNA as the core genetic material. An added bonus was the potential for rapid replication of compact DNA genomes. The full potential of DNA genomes was mostly realized after divergence of archaea and bacteria.

As a survival and replication strategy, eukaryotes opted for more complex genomes, more complex chromosomes, more complex metabolism, more complex gene regulation, more complex signaling, and generally more complex multicellular organizations than archaea and bacteria.

The ribosome, however, is an ancient stalwart of living systems that enabled the inception of protein evolution to compete with and complement RNA evolution. The tree of life built based on ribosome RNA sequences is one of the more reliable trees, but organisms and ribosomal RNA are not the only currencies of life on earth, and other metrics produce different and to an extent seemingly incompatible trees.

Dumb polymerization underlies RNA coding, ribosome-dependent protein synthesis/folding/structure/function/dynamics, and DNA coding. Life is made up of dumb polymers, and, in biology, smart is often dumb and dumb may be smart or, at least, survivable. Evolutionary purpose develops in retrospect, defined by purifying selection, killing unsuccessful designs.

Chapter 23

The RNA-Protein World

Once the ribosome or proto-ribosome evolves, we enter the RNA-protein world.

To a molecular biologist, the RNA-protein world is startlingly familiar.

Life of this ancient time had RNA genomes, ribosomes or proto-ribosomes (ribosomal RNA), ribozymes, transfer RNA, diverse RNA modifications, messenger RNA, many proteins, and primitive metabolism.

Coevolutionary forces develop competition between ribozymes and proteins to support catalytic functions. As protein cofactors have invaded the ribosome, protein cofactors evolved to invade each ribozyme. Surprisingly, however, protein cofactors that surround and inundate ribosomal RNA failed to take over the ribosome's fundamental RNA-based catalytic or clustering function in protein synthesis. So, for peptide bond formation, the ribosome remains a ribozyme studded with protein cofactors. Gazing on the ribosome, ancient competition and coevolution of proteins and RNA remain apparent (Fig. 19.1). In some measure, therefore, the RNA-protein world can still be seen.

The RIFT barrel is a compact six β-sheet barrel with all antiparallel β-sheets. This ancient barrel is the mother of the double-Ψ–β-barrel that I have described in detail (Chapter 17).

The essential process of RNA synthesis was taken over from presumed self-replicating gene-ribozymes by two double-Ψ–β-barrel-type RNA-template-dependent RNA polymerases with a bridge helix, a trigger loop and a 2-Mg^{2+} catalytic mechanism. These enzymes and their descendants endure in very recognizable forms today (Chapters 17 and 18).

As proteins began to take over primitive metabolism, which was initially supported by ribozymes, $(\beta-\alpha)_n$ repeat proteins became a major life force. TIM barrels $(\beta-\alpha)_8$ and Rossmann folds $(\beta-\alpha)_8$ were major scaffolds on which to build enzymology (TIM barrels) and redox energy transduction (Rossmann folds and TIM barrels). The RNA-protein world probably had glycolysis, the citric acid cycle, the glyoxylate cycle, and many other catalytic competencies. Metabolism began to assume a strikingly modern cast. Metabolism utilized redox energy transduction, so battery-powered metabolism was in place. NAD, FAD, and FMN oxidoreductase enzymes were established in the most ancient of times.

What about chemical energy transduction?

$(\beta-\alpha)_5$ Walker motif kinases, ABC ATPases (complex $\beta-\alpha$ repeat folds), Swi-Snf ATPases (two $(\beta-\alpha)_6$ domain proteins), and other $(\beta-\alpha)_n$ repeats

Evolution since Coding. http://dx.doi.org/10.1016/B978-0-12-813033-9.00023-8

evolved. So ATPases, GTPases, and kinases were born in the RNA-protein world, and all of these remain as $(\beta-\alpha)_n$ repeat folds today. In addition to battery driven (redox) energy transduction, these enzymes form the basis for chemical energy transduction that allows living systems to run a chemical currency of ATP and GTP hydrolysis.

$(\beta-\alpha)_n$ repeat proteins also catalyze the charging of tRNAs with their cognate amino acids, which is an ancient and essential function of the RNA-protein world.

Frighteningly, we have reconstructed much of the ancient protein component of the RNA-protein world using a few six and eight β-sheet barrels and some derived $(\beta-\alpha)_n$ repeats. Of course, there are other parts to this story, but the concept is the key. Generate a few simple repeat structures using a few simple protein motifs. Generate different folds. Make many gene copies. Diverge enzymes to generate many catalytic specificities. Startlingly, once established, these simple and repetitive protein folds become immortal and last for ~4 billion years.

RNA genomes in the RNA-protein world are more volatile and less streamlined than DNA genomes. RNA segments tend to be small. RNA is chemically somewhat unstable, for which you may blame the 2′-OH in RNA and single strands. In the RNA-protein world, RNA genomes were made of many segments with each RNA in many copies. Genes were more independent of one another and individual gene copies were in competition. Therefore, the RNA genome world was a rich crucible for generating protein and enzyme diversity. RNA genomes were tangles of semiindependent replication units. Enzymatic RNA ligation could be used to link two identical RNA molecules to generate a dimeric repeat. Repeating the process generated additional motif repeats. Replication errors were common, including duplications and multimerizations. Looking at the frequency of repeating motifs in protein folds today, it is clear that in the ancient world mechanisms for generation of repeating sequences and motifs were a primary driving force toward genetic and protein complexity.

Despite its strangeness, the RNA-protein world is very recognizable to a molecular biologist.

The RNA genome world is similar to many viruses with RNA genomes. Many of the same rules for volatility, mutation, and evolution apply.

Imagine encapsulated seas of multicopy, semiselfish, and semiautonomous RNA elements mixed with simple LEGO (Trademark) peptides, which in large part were $(\beta-\alpha)_n$ protein fragments. Understanding structural and functional interactions of $(\beta-\alpha)_{2,4,8}$ repeats and cradle-loop barrels with single and double-stranded RNAs would offer many new surprising insights into the strange RNA-protein world of ~4 billion years ago.

From $(\beta-\alpha)_n$ repeat proteins, cradle-loop barrels, ribosomes, riboswitches, and ribozymes, the RNA-protein world can be glimpsed in today's world, perhaps even with more painful clarity than Sarah Palin descries/decries Russia. Oddly, in 2017, the Republicans are completely in tune with fascist Russia.

Chapter 24

Transfer RNA

Translation mechanisms are too ornate (Fig. 19.1), so how did they evolve?

Here, it should be noted that, although the ribosome may appear to represent something resembling irreducible complexity, the ribosome does not appear to represent intelligent design. The ribosome is a molecular horror that appears evolved.

Because translation is such a central function of information processing in cells, multiple systems coevolve with the ribosome, and simplification of the ribosome structure becomes progressively more difficult. Transcription systems are similar. Hysteresis of core functions and coevolution of multiple codependent functions tend to limit subsequent evolutionary drift. Too many peripherals depend too heavily on core translation and transcription systems.

The central molecule in translation is transfer RNA (tRNA) (Fig. 24.1). tRNA is a relatively rigid structure with a rigid anticodon loop and a rigid and sharp "elbow." The anticodon loop and the T-loop of tRNA are similar in sequence and structure. The 7 nucleotide anticodon loop is rigid because it forms a sharp U-turn between anticodon loop residues 2 and 3. Loop residues 3–5 are the anticodon positions. Because the mRNA has the opposite polarity to the anticodon, loop nucleotide 3 is the "wobble" (corresponding to the third and somewhat ambiguous position of the codon in mRNA) position of the anticodon. In the anticodon loop, loop nucleotides 3–7 form an even stack of bases. At the elbow, the T-loop interacts with the D-loop. Specifically, D-loop G19 intercalates into the U-turn loop increasing the spacing between T-loop nucleotides 4 and 5, elevating nucleotide 5 to fill the loop and forcing nucleotides 6 and 7 out of the loop. Differences in structure between the homologous anticodon loop and the T-loop, therefore, are because the T-loop interacts with the D-loop to form the rigid elbow. The aminoacylated 3'-CCA ends of the tRNAs are clustered together within the peptidyl transferase center of the ribosome, where peptide bonds are formed.

Because tRNA is the molecular archetype around which translation systems evolved, to make sense out of translation mechanisms, start with tRNA and work outward.

Evolution is the concept underlying biology; therefore to understand overly ornate translation mechanisms, assume a simple stepwise evolutionary mechanism, and see where that gets you.

Evolution since Coding. http://dx.doi.org/10.1016/B978-0-12-813033-9.00024-X

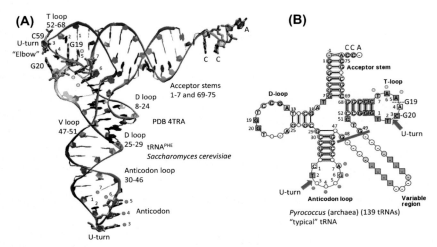

FIGURE 24.1 Cloverleaf transfer RNA (tRNA). (A) tRNA structure. The 5′- and 3′-acceptor stems (As) are green; the D-loop microhelix is pink; the anticodon loop and the T-loop are red; the last 5 nucleotides of the D-loop and the V-loop are cyan; the 3′-CCA is yellow. Some important residues are drawn for chemistry. (B) tRNA cloverleaf diagram. 139 *Pyrococcus* tRNAs (from 3 archaeal species) were collected from the tRNA database (http://trna.bioinf.uni-leipzig.de/) and used to generate a "typical" tRNA cloverleaf diagram (similar to a consensus sequence). Numbering is based on a 75 nucleotide tRNA core. *Red arrows* indicate sharp U-turns. *Blue dots* indicate anticodon positions. *Orange dots* indicate stacked bases within the anticodon loop. *Yellow dots* indicate elevated (loop position 5), and flipped out (loop positions 6 and 7) bases in the T-loop.

It is difficult to prove a particular ancient evolutionary mechanism, but, perhaps, this is not the central issue. Without considering evolution, no reasonable working model for genesis can be devised; therefore no reasonable understanding is possible. With consideration of evolution, working models for evolution come alive. These models need not be absolutely true in every detail to be informative. Working models generate testable hypotheses to obtain greater depth of understanding.

So, how did translation evolve?

How did tRNA evolve and how could one know?

tRNA must be >3.5 billion years old, at a minimum. So with ongoing evolutionary pressure and mutation, how could a mechanism for tRNA evolution be imagined >3.5 billion years after the fact? Would not the sequence/structure mutate too much in all that time for the initial sequence patterns to erode beyond recognition? Because ancient patterns are recognizable for proteins, perhaps ancient patterns remain recognizable for tRNA.

Based on its secondary structure, tRNA is described as a "cloverleaf" (Fig. 24.1B). In 3-D, however, the molecule is "L-shaped" (Fig. 24.1A). The D-loop and the T-loop interact to form the "elbow" of the L. Overall, tRNA is a rigid structure to support its role as a translation adapter. In the cloverleaf diagram, the anticodon loop and the T-loop closely resemble each other.

tRNA is considered an "adapter." One end of tRNA presents a complementary sequence to the mRNA (messenger RNA) (this is the anticodon loop of tRNA matched to a codon in mRNA). The other end of the tRNA attaches to an amino acid that can be transferred to a growing polypeptide chain: a protein (this is the 3′-CCA end of the tRNA) (Fig. 24.1). In this way, tRNA forms the adapter molecule linking the RNA sequence in the mRNA to the amino acid sequence in a polypeptide chain. "Translation," therefore, changes the alphabet from RNA sequence (A, G, C, U) in mRNA to protein sequence (20 amino acids).

Francis Crick very cleverly proposed the adapter hypothesis before much of the molecular biology was known. When Francis Crick saw the tRNA cloverleaf structure, he thought it was too large to be the adapter he had imagined. So Crick wanted a smaller adapter (i.e., ~25 nucleotides, according to one of his studies). Perhaps, in support of Crick's intuition, a smaller adapter can be identified, and cloverleaf tRNA is part of a second generation coding mechanism. I shall argue here, based on the tRNA structure, that cloverleaf tRNA could be a third generation mRNA→protein coding scheme. Fig. 24.2 shows multiple generations of proto-tRNAs and acceptor stems that may have preceded cloverleaf tRNA. Each of these pieces is cut directly from the cloverleaf structure as if the tRNA cloverleaf is a fossil conglomerate of more ancient RNA pieces. The RNA-protein world appears to precede the cloverleaf tRNA world and LUCA. The cloverleaf tRNA world was presaged by a proto-tRNA minihelix world and possibly, before that, a proto-tRNA microhelix world. Either the microhelix (17 nucleotides + 3′-CCA) or the minihelix (31 nucleotides + 3′-CCA) is similar

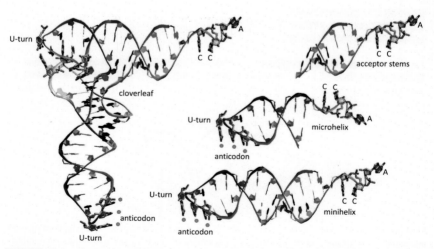

FIGURE 24.2 Multiple posited generations of translation adaptor embedded in cloverleaf tRNA: 75 nucleotide cloverleaf (+3′-CCA), 31 nucleotide minihelix (+3′-CCA) and 17 nucleotide microhelix (+3′-CCA). Acceptor stems could also have participated in peptide joining. Colors are as in Fig. 24.1A.

to a Francis Crick adaptor. The 31 nucleotide minihelix is a 17 nucleotide micro-helix with an acceptor stem (2×7 nucleotides) (31 nucleotides + 3'-CCA). The acceptor stem is recognized by the enzyme (amino acyl tRNA synthetase) that adds an amino acid to the tRNA 3'-CCA end. So, the acceptor stem is an addition to a microhelix to ensure the accuracy of amino acid attachment to tRNA to improve the accuracy of protein synthesis.

So, how did the cloverleaf tRNA structure evolve?

It turns out that tRNA structure (75 nucleotides + 3'-CCA) and its evolution can be solved as a surprisingly simple puzzle down to the last nucleotide. This model was designed by Robert Root-Bernstein and me (Michigan State University).

Cloverleaf tRNA is made up from ligation of three 31 nucleotide minihelices (Fig. 24.3).

$3 \times 31 = 93$ nucleotides, so this is too long, by 18 nucleotides.

But with two internal nine nucleotide symmetrical deletions within the joined acceptor stems, $93 - 18 = 75$, which is the correct answer. So, how do we get there?

First of all, what is a minihelix, and why 31 nucleotides?

A 31 nucleotide proto-tRNA minihelix structure is shown in Fig. 24.2. A 17 nucleotide proto-tRNA microhelix is shown, lacking an acceptor stem (2×7 nucleotides: $31 - 14 = 17$ nucleotides). The microhelix or the minihelix looks like the Crick adapter. The minihelix looks like a microhelix linked to an acceptor stem for greater accuracy in amino acid attachment.

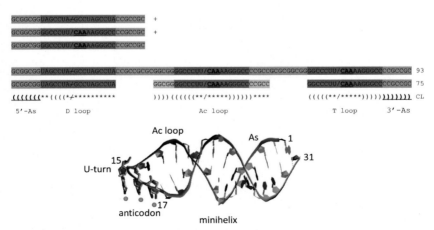

FIGURE 24.3 A model for the genesis of the 75 nucleotide cloverleaf tRNA by ligation of 3–31 nucleotide proto-tRNA minihelics. Three 31 nucleotide minihelices (17 nucleotide microhelices (pink or red) + acceptor stems (green)) are ligated together followed by symmetrical 9 nucleotide deletions within ligated acceptor stems. Cloverleaf tRNA secondary structure is indicated ("()" indicates base pairing; * indicates loop; / indicates an RNA U-turn; / indicates no U-turn). An image of a 31 nucleotide minihelix lacking 3'-CCA is shown below the diagram.

It should be noted here that the CCA 3'-end, where the amino acid is attached, is added to the tRNA enzymatically in many archaeal and eukaryotic species. One thing this means is that the CCA end is a separate issue from evolution of the core cloverleaf tRNA (75 nucleotides).

OK. So back to our model. We take three 31 nucleotide minihelices and ligate them together (Fig. 24.3). Within the two internally joined 14 nucleotide CG-rich acceptor stems, we generate symmetrical 9 nucleotide deletions, leaving 5 nucleotide remnants of acceptor stems on both sides touching the anticodon loop 17 nucleotide microhelix. And *wallah*! We get cloverleaf tRNA.

So, why do we think this is a good model?

To understand tRNA structure, one must understand acceptor stems, the D-loop, the anticodon loop, the V-loop, and the T-loop (Fig. 24.1). The basic V-loop is just 5 nucleotides, so, for evolutionary modeling, only 5 V-loop nucleotides need be considered. To start working the puzzle, notice that the anticodon loop and the T-loop are related to one another by sequence and structure (Figs. 24.1, 24.4, and 24.5). These sequence and structural similarities make the anticodon loop and the T-loop genetic relatives (homologs). Fig. 24.4 presents the sequence comparison using sequence logos. Sequence logos are a reasonably intuitive visual demonstration of sequence relatedness that uses the colors and size of letters to indicate the prevalence of a particular nucleotide (here shown as DNA sequence). From these and other data, we infer that the anticodon loop and the T-loop have an ancestral sequence very close

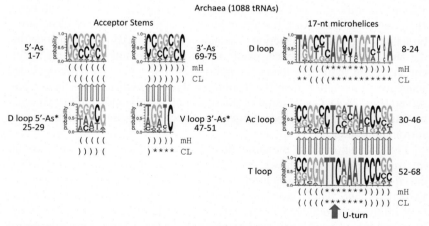

FIGURE 24.4 Sequence logos of 1088 archaeal tRNAs. Sequence logos are a means to visualize the similarities and differences of aligned sequences by viewing the fraction of a nucleotide base (shown here as DNA) at each position. According to the tRNA evolution model, the 75 nucleotide tRNA core can be divided into the 5'-acceptor stem (1–7), the 3'-acceptor stem (69–75), a 5'-acceptor stem remnant (D-loop 25–29), a 3'-acceptor stem remnant (V-loop 47–51) and 3–17 nucleotide microhelices (D-loop 8–24; anticodon loop 30–46; T-loop 52–68). *Yellow arrows* indicate inferred homologous positions. Secondary structures are indicated. CL indicates cloverleaf tRNA folding; mH indicates minihelix and microhelix folding.

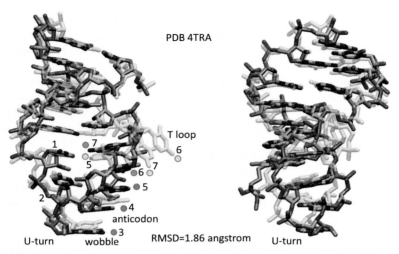

FIGURE 24.5 The anticodon loop and the T-loop are homologous structures. Overlaid anticodon loop (colored by chemistry) and T-loop (transparent yellow). Two views are shown. *Blue dots* indicate anticodon positions. RMSD (root-mean-square deviation) indicates the similarity of the structures.

to CCGGGUU/CAAAACCCGG [/ indicates the position of the U-turn (Figs. 24.1 and 24.2)]. Fig. 24.5 shows, furthermore, that the anticodon loop and the T-loop structures can be overlaid precisely showing that, in addition to being close sequence homologs, the anticodon loop and the T-loop are close structural homologs. From inspection of sequence logos, one can also conclude that the anticodon loop sequence, which needs to accommodate multiple different anticodon sequences to encode multiple amino acids, is a more degraded sequence in evolution than the T-loop sequence, which is much better preserved (Fig. 24.4). Coevolution of the surrounding anticodon loop is expected because changing the anticodon sequence is bound to drive the evolution of the surrounding loop.

Fig. 24.5 shows that the anticodon loop and the T-loop are very similar in structure. Fig. 24.6 shows that, where the structures are slightly different, it is because the T-loop binds to the D-loop to form the fairly rigid cloverleaf "elbow." The T-loop varies from the anticodon loop because D-loop G19 intercalates between T-loop positions 4 and 5. In the anticodon loop, loop residues 3 to 7 stack on one another. In the T-loop, intercalation of D-loop G19 pushes loop bases 4 and 5 apart, causing loop base 5 to fill the loop and, therefore, causing loop bases 6 and 7 to flip out of the T-loop (Figs. 24.1, 24.5, and 24.6). If the tRNA cloverleaf were too flexible, it would be awkward to position the anticodon end relative to the amino-acylated end, as required for the translation adapter. So tRNA has a sharp, rigid elbow, and a rigid anticodon loop.

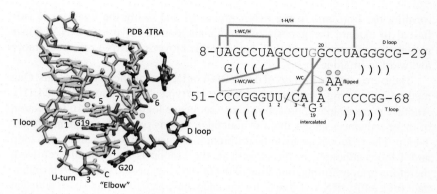

FIGURE 24.6 Structural differences between the anticodon loop and the T-loop are due to T-loop–D-loop contacts. The T-loop is yellow. The D-loop is colored by chemistry. D-loop G19 intercalates between T-loop nucleotides 4 and 5, elevating nucleotide 5 in the loop (*yellow dot*) and flipping out nucleotides 6 and 7 (*yellow dots*). The schematic indicates D-loop–T-loop interactions. Non-Watson-Crick base pairing interactions are indicated. *H*, Hoogsteen; *t*, trans; *WC*, Watson-Crick. *Thin blue lines* indicate weak interactions.

The T-loop is linked to a 7 nucleotide acceptor stem, indicating that the T-loop was once a 31 nucleotide minihelix (Figs. 24.1 and 24.2).

The modern cloverleaf tRNA structure includes a 31 nucleotide minihelix, with a 17 nucleotide T-loop microhelix linked to a 7 nucleotide paired acceptor stem (Figs. 24.1 and 24.2). Because this minihelix structure is part of cloverleaf tRNA, 31 nucleotide minihelices are posited to have existed before cloverleaf tRNA. Furthermore, if the T-loop is derived from a 31 nucleotide minihelix, the anticodon loop is posited to be derived from a 31 nucleotide minihelix because the anticodon loop is a homolog of the T-loop. Because these sequences flank the anticodon loop, the 3′-end of the D-loop [5 nucleotides; 25–29 (Fig. 24.1)] and the V-loop (5 nucleotides; 47–51) might be expected to be remnants of acceptor stems.

But the D-loop is also linked to a 7 nucleotide acceptor stem. Therefore we posit that the D-loop was also derived from a 31 nucleotide minihelix. So, now we think that three 31 nucleotide minihelices, two of which were identical, were ligated to form a 93-nucleotide precursor to the cloverleaf tRNA. Now we must delete 18 nucleotides to obtain the cloverleaf tRNA sequence and structure (Fig. 24.3).

Let us link up three 31 nucleotide minihelix sequences (Fig. 24.3) and see how we are doing with our model.

This is pretty much like making model airplanes as a kid.

So, make the two 9 nucleotide deletions symmetrically, as indicated in Fig. 24.3. If you do this, you generate the entire 75-nucleotide tRNA cloverleaf molecule.

The precise biochemical mechanism for making the symmetrical two 9 nucleotide deletions is not currently known, although RNases can be identified that make similar cuts in RNA. RNA ligases link RNAs 5′ to 3′, so forming the four cuts in the tRNA precursor and doing the ligations is sufficient to explain the genesis of the 75 nucleotide cloverleaf tRNA. Other models for construction of the 75 nucleotide cloverleaf tRNA are much more awkward to imagine.

There are a couple of ways of doing RNA or DNA sequence comparisons. One method is by using consensus sequences or "typical" sequences (Fig. 24.1B). Another more visible approach is to use sequence logos (Fig. 24.4). Inspection of these sequences shows that the anticodon loop is related by sequence to the T-loop. The D-loop appears to be different in sequence from the anticodon loop and T-loop, although, in the cloverleaf, the D-loop is substantially refolded; therefore, after refolding, mutation of the D-loop is expected to support the alternate fold. Most tRNA D-loops, for instance, have deletions, but some archaeal tRNA D-loops remain intact. So far as I could tell, inspecting many tRNA from many species, very few bacterial or eukaryotic D-loops remain without deletions. Among other considerations, this means only archaeal tRNAs can easily be used to generate the model for tRNA evolution. Archaea are an old domain of life and, for tRNA structure and sequence, archaea is the domain most similar to LUCA (the last universal common ancestor). Here, for convenience, we compare DNA sequences encoding tRNA using "typical" sequences and sequence logos.

As noted above, the anticodon loop and the T-loop are very similar in sequence and maybe, initially, were identical. These are homologous sequences and structures. The T-loop appears to be derived from a CAA leucine encoding minihelix. The anticodon loops must now encode all 20 amino acids, so the anticodon sequence and its surrounding loop is now highly variable. Initially, the anticodon loop may have been CAA specifying leucine similar or identical to the T-loop. Alternatively, the T-loop wobble nucleotide could have mutated to become C so that it could form a Watson-Crick base pair to G20 of the D-loop (Figs. 24.1 and 24.6).

The anticodon loop and the T-loop appear to be derived from the ancestral sequence ~ CCGGGUU/**<u>CAA</u>**AACCCGG. The / indicates the placement of a U-turn, a sharp U-shape turn in the RNA backbone, right after a U in the tRNA sequence. The anticodon sequence is bold and underlined. After the advent of cloverleaf tRNA, all anticodon loops have this basic shape and form. These are 7 nucleotide loops on 2×5 nucleotide paired stems (17 nucleotides total). Between loop positions 2 and 3 is the sharp and necessary U-turn. In tRNA, the U-turn essentially always occurs after a U in the loop sequence. The U-turn appears essential to present a 3 nucleotide anticodon sequence to mRNA for translation on a ribosome. In tRNA, the anticodon loop with its sharp U-turn is quite a rigid loop. An evidence of its rigidity is that a tRNA that is free in solution is very similar in conformation to a tRNA bound to mRNA on the ribosome (Figs. 24.7 and 24.8). Because an adapter must present nucleotides at one end and an amino acid or peptide chain at the other, structural rigidity is useful in an adapter. Otherwise, aminoacylated tRNA

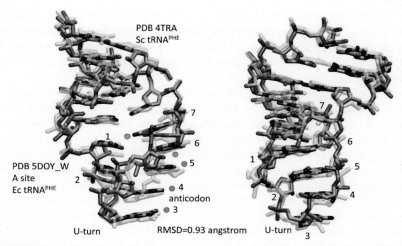

PDB 4TRA
Sc tRNA^PHE

7
1
6
5
PDB 5DOY_W
A site
Ec tRNA^PHE 2
4
anticodon
3

U-turn   RMSD=0.93 angstrom   U-turn

7
1
6
5
2
4
3

FIGURE 24.7 Free tRNA (transparent yellow) has a similar structure to tRNA bound to mRNA on a ribosome (colored for chemistry).

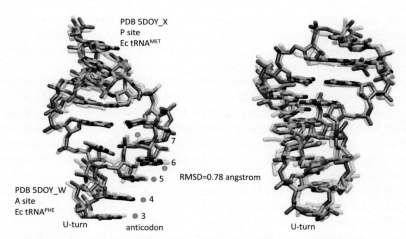

PDB 5DOY_X
P site
Ec tRNA^MET

7
6
5   RMSD=0.78 angstrom
PDB 5DOY_W
A site
Ec tRNA^PHE   4
U-turn   3
          anticodon   U-turn

FIGURE 24.8 tRNAs bound to mRNA on a ribosome are very similar in structure.

ends might get lost on the ribosome inhibiting translation. The 17 nucleotide CCGGGUU/**CAA**AACCCGG microhelix with a U-turn in the 7 nucleotide loop, therefore, as a rigid adapter, assumes significant biological value from the ancient proto-tRNA world to the present day. Significantly, this 7 nucleotide loop appears essential to present a 3 nucleotide anticodon in a translation adapter. Following this reasoning, standardization to a 3 nucleotide translation code may have happened around the time of LUCA when cloverleaf tRNA appears to have evolved.

The D-loop anticodon sequence is GCC, which might specify glycine. The difficulty with this assignment, however, is that the D-loop does not appear to form the appropriate fold to present a GCC 3 nucleotide anticodon. So, making sense out of the D-loop 17 nucleotide microhelix requires some consideration.

First of all, the D-loop 17 nucleotide microhelix appears to be derived from a truncated $(UAGCC)_4$ repeat: **UAGCC**UAGCCUAGCCUA. From sequence logos, the archaeal D-loop microhelix has the likely sequence UAGCCUAGCCU**GG**UCUA, which is very close to the posited repeat sequence [varying in only two positions (bold and underlined)]. So the D-loop 17 nucleotide microhelix is very different in sequence from the anticodon and T-loop microhelices. Furthermore, it does not appear that UAGCCUA/**GCC**UAGCCUA can form a U-turn to present GCC as an anticodon. So far as I know, U-turns never occur after A, so this is a very bad sequence to form a 7 nucleotide anticodon loop. This indicates that the D-loop is another class of 17 nucleotide microhelix from CCGGGUU/**CAA**AACCCGG. If the D-loop microhelix had a role as a translation adapter before cloverleaf tRNA, therefore, it probably did not present a 3 nucleotide anticodon. This argument suggests that the 3 nucleotide genetic code was standardized close to LUCA, with the advent of cloverleaf tRNA. Unfortunately, for our current understanding, it appears that (at most) 2 microhelix sequences survived the transition to the cloverleaf tRNA world, and one of those sequences, the D-loop, is problematic as a translation adapter presenting a three nucleotide anticodon.

The two 5 nucleotide sequences that flank the anticodon 17 nucleotide microhelix include parts of the D-loop (25–29) and the V-loop (47–51). These sequences resemble acceptor stem sequences (Fig. 24.4 and other data), as predicted from the model. Because acceptor stems are 7 nucleotides long, it is very difficult to think of a mechanism to generate these two 5 nucleotide acceptor stem fragments except by ligation and internal symmetric deletion, in support of the model (Fig. 24.3).

As with the D-loop microhelix truncated $(UAGCC)_4$ repeat, acceptor stems appear to be generated from a truncated $(GCG)_3$ repeat (Fig. 24.1, right panel; Fig. 24.4). The archaeal 5′-acceptor stem sequence is close to 1-**GCG**GCGG-7. The ancestral sequence of the 5′-acceptor stem remnant (close to 25-GGCGG-29) appears to correspond to 5′ acceptor stem 3-GGCGG-7. The 3′-acceptor stem (derived from 69-C**CGC**CGC-75) and the 3′-acceptor stem remnant (derived from 47-CCGCC-51) are complementary sequences to 5′-acceptor stems, generated by complementary replication. So, according to this analysis, 41 out of 75 nucleotides of cloverleaf tRNA was derived from simple contiguous GCG and UAGCC repeating RNA sequences, and, remarkably, in archaeal sequences, a strong print remains after >3.5 billion years of these initial repeats. The rest of the cloverleaf tRNA is a 2X noncontiguous repeat of CCGGGUU/**CAA**AACCCGG microhelices [2×17 nucleotides (anticodon loop (30–46) and T-loop (52–68))], so every single nucleotide in cloverleaf tRNA is accounted for by a surprisingly simple model of evolution, and 75 out of 75 nucleotides in the tRNA core arise from a repeating sequence of 3 (GCG), 5 (UAGCC), and 17 (~CCGGGUU/CAAAACCCGG) nucleotides.

It appears that the proto-tRNA minihelix world, which preceded the cloverleaf tRNA world, included a set of minihelices encoding several amino acids.

Before the proto-tRNA minihelix world, one can imagine a 17 nucleotide microhelix proto-tRNA RNA-protein world, for instance, before amino acyl transferase enzymes utilized an acceptor stem to help specify the amino acid to be joined to the proto-tRNA. Very likely, the 17 nucleotide microhelix world (if such existed) encoded fewer amino acids than the 31 nucleotide minihelix world.

So, many innovations, including cloverleaf tRNA, occurred at about the time of LUCA, which is the first cellular organism. Cellular translation systems to synthesize protein were fundamentally changed at the time of LUCA with the advent of cloverleaf tRNA, but have not changed very much since. The 31 nucleotide proto-tRNA minihelices and 17 nucleotide proto-tRNA microhelices appear to remain embedded in the tRNA cloverleaf structure (Figs. 24.1 and 24.2), which itself appears to be >3.5 billion years old.

Furthermore, many core molecules, including tRNA, remain substantially unchanged since LUCA.

If 17 nucleotide proto-tRNA microhelices and 31-nucleotide proto-tRNA minihelices supported mRNA-directed protein synthesis, how might these primitive translation systems have worked? Well, here is an idea.

Despite its complexity (Fig. 19.1), the ribosome does a small number of simple jobs. The ribosome is a scaffold on which mRNA-directed protein synthesis takes place. On the small ribosomal subunit, a segment of mRNA is loosely restrained, so that tRNA anticodons can be paired to mRNA. The small ribosomal subunit (in archaea and bacteria) is constructed around the 16S rRNA (16S is a measure of RNA size). On the large ribosomal subunit, two tRNAs are involved in peptidyl transfer: the formation of peptide bonds in protein synthesis. One tRNA with a bound amino acid occupies the ribosome "A" site. One tRNA with a bound polypeptide chain occupies the ribosome "P" site. Bringing two aminoacylated tRNAs into close proximity within the peptidyl transferase center of the large ribosomal subunit allows peptide bond formation. The peptidyl transferase center mostly gathers and positions the two aminoacylated tRNAs. In bacteria and archaea, the large ribosomal subunit is built around the 23S rRNA, and domain 5 of the 23S rRNA comprises the peptidyl transferase center. After peptidyl transfer, the tRNA in the "A" site holds the polypeptide chain, which is now one amino acid longer in length. The mRNA moves over one step (translocates), so the tRNA holding the polypeptide chain is now in the "P" site. Another aminoacylated tRNA occupies the "A" site, and the process continues until the polypeptide chain is terminated.

So, the ribosome is a simple machine and scaffold to make proteins in a mRNA-directed fashion. Cloverleaf tRNA spans the two ribosomal subunits. The two anticodon loops of the two tRNAs bind two adjacent-mRNA codons on the small ribosomal subunit. The two 3'-CCA-amino acid ends of the tRNAs are brought close together for peptidyl transfer within the peptidyl transferase center of the large ribosomal subunit. The bond is formed and the mRNA and tRNAs translocate.

Cloverleaf tRNA can span the distance between the mRNA held on the small ribosomal subunit and the peptidyl transferase center on the large ribosomal subunit, but microhelix proto-tRNA or minihelix proto-tRNA cannot. So, the proto-ribosome was probably a single subunit, similar to the small ribosomal subunit of the cellular ribosome, and the peptidyl transferase center was probably separate and mobile. The proto-ribosome, therefore, is imagined as a scaffold to hold and translocate the mRNA, and the proto-peptidyl transferase center is considered to be a mobile piece to cluster the amino-acylated proto-tRNA 3'-CCA-amino acid ends. This view indicates that the 16S rRNA is older than the 23S rRNA, which indeed it is, based on its higher evolutionary conservation.

So, complexity aside, the concept is simple. tRNAs are adapters with two ends. One end is the anticodon loop that binds the codon in mRNA. The other end is the 3'-CCA-amino acid end. Bring two tRNAs (or proto-tRNAs) together within a peptidyl transferase center, and a peptide bond is formed. The proto-ribosome is primarily a scaffold to hold the mRNA in place and to allow for mRNA movement (translocation). The proto-ribosome appears to be a single subunit including proto-16S rRNA with a decoding center to hold the mRNA. The proto-ribosome probably requires a mobile peptidyl transferase center.

According to a tRNA-centric view of ribosome evolution, bringing two aminoacylated tRNAs close together within a peptidyl transferase center appears to be sufficient to achieve polypeptide bond formation. Thus to grasp mRNA-dependent translation, all you need is a proto-ribosome scaffold to retain the mRNA to position the aminoacylated tRNAs. According to this view, tRNAs are the central function of the peptidyl transfer reaction, and ornate translation systems evolved around proto-tRNA adapters.

Other features of ribosomes that facilitate translocation, initiation, and termination are later evolutionary add-ons and improvements.

According to this mode of thinking, the ornate two subunit cellular ribosome, much of rRNA, cloverleaf tRNA, and standardization to a 3 nucleotide genetic code were major codependent evolutionary innovations at about the time of LUCA.

Furthermore, this is an example of the apparent explosive evolution of cellular, DNA-based life on earth. Proto-tRNA microhelices and minihelices were short-lived species that were very quickly replaced by cloverleaf tRNAs by the process described in this chapter. As soon as 31 nucleotide minihelices with U-turn 7 nucleotide loops evolved, a pathway opened to evolve cloverleaf tRNA, which could evolve to specify 20 amino acids utilizing a 3 nucleotide genetic code. The radical cloverleaf tRNA innovation drove the evolution of the modern cellular ribosome and ornate cellular translation systems. The cloverleaf tRNA revolution appears to correlate with the evolution of the first, cellular, DNA-based life.

Consider this a fairly straightforward and simple working model for evolution of cloverleaf tRNA from proto-tRNA microhelices and minihelices. To me, the amazing thing is that, after >3.5 billion years (the age of cloverleaf tRNA), such a model can be imagined and designed. The model provides conceptual insight into the evolution of the overly ornate, two subunit cellular ribosome.

The model presented for evolution of cloverleaf tRNA is quite straightforward. Cloverleaf tRNA was evolved from ligation of three 31 nucleotide proto-tRNA minihelices and two symmetrical 9 nucleotide deletions within joined acceptor stems (Fig. 24.3). This is more like a puzzle solution or a mathematical proof than an evolutionary working model. Remarkably, this model describes evolution of 75 out of 75 nucleotides in core cloverleaf tRNA. This model conceptually links evolution, genetics, biochemistry, and structure in a very convincing and satisfying way. This model for tRNA evolution is very similar to models for protein evolution, involving repetition and joining of simple motifs, as described in other chapters. In very many cases, biological complexity, therefore, develops through iteration of simple motifs to join repeating sequences. The reason repetition of motifs is important in evolution appears to be that iteration is sometimes the fastest initial evolutionary route to complexity, solubility, and structure. In many core macromolecules, therefore, >3.5 billion years after LUCA, iterated patterns remain recognizable in sequence and structure.

GENERATION OF SEQUENCE REPEATS

In the RNA-protein world, there is a simple mechanism to generate repeating RNA sequences. Significantly, RNA genes can potentially maintain some independence from other RNA genes, independence that is largely lost after LUCA and the advent of rapidly replicating DNA genomes. Simply stated, RNA genes can exist, replicate, and ligate in isolated colonies to generate repeating sequences. Replication of RNAs is thought to have occurred by ligation of snap-back primers. For instance, a 17 nucleotide microhelix or a 31 nucleotide minihelix lacking 3'-CCA can act as a snap-back primer when ligated to other RNAs. Any RNA that starts with a double-stranded stem can act as a snap-back primer. Short RNAs, furthermore, must ligate together to be replicated, so, in an isolated RNA colony, RNA repeats are naturally generated. Because RNAs are linked together by ligation for replication, they must also be separated into useful segments by RNA processing enzymes, some of which are ribozymes, such as ribonuclease P. Furthermore, ligation of many RNAs together generates molecules of the size and complexity of ribonuclease P, ribozymes, and ribosomal RNAs.

Based on success with the cloverleaf tRNA evolution model, similar attempts have been made to describe a simple pathway for evolution of ribosomal RNA. I guess that such a model can be devised, but, as of 2017, no such model is yet apparent (to me).

"INTELLECTUAL PROPERTY" IN BIOLOGY AND EVOLUTION

tRNA, one of the most central information processing molecules in living systems, is built of dumb GCG, UAGCC, and ~CCGGGUU/**CAA**AACCCGG repeats and nothing else, except 3'-CCA. I have argued that 75 out of 75 nucleotides of the tRNA core were built from these 3 repeating sequences. Because archaeal tRNA has conserved these repeats, astoundingly, a record of the transition from the inanimate repeating polymer world to a world that includes enduring biological value is preserved in the cloverleaf structure. This makes tRNA into an amazing molecular fossil that records an ancient transition between the inanimate and the animate. As such, tRNA is a little strange and frightening.

TRY THIS AT HOME

Prove to yourself that the anticodon loop and the T-loop are structural homologs/paralogs. This is not difficult, but it requires some manipulations of PDB files. For reasons I do not completely understand, the simple procedure described here will not work with every tRNA structure, but, before the author lost patience, the approach was made to work with PDBs 1YFG, 4TRA, and 2AKE. Of course, there are other strategies for doing these structural alignments, but I am not very good with computers, so I would not try to explain. In this case, the first thing to do is to isolate the anticodon loop and the T-loop as separate PDB files (with the same number of nucleotides). The next step is to align the two PDB files using the program Visual Molecular Dynamics and the RMSD calculator under Analysis. The program Pymol is convenient for isolating the anticodon loop and the T-loop as independent PDB files. If you have the patience, this can also be done by editing PDB files using NotePad. If all goes well, you can align the backbones of the two loops within ~1.9 Å RMSD (root–mean-square deviation). The anticodon loop and the T-loop really are homologs/paralogs (Fig. 24.5). The VMD alignment tool (the RMSD calculator) rebels if it is trying to align different numbers of atoms. If things become too confusing, try starting over from scratch. Computers are frustrating. VMD, I think, will only do one alignment in a single session. So, if you get a funny alignment with VMD, you must start again, and you will probably need to modify your approach.

NOW TRY THIS

Go to the tRNA database (http://trna.bioinf.uni-leipzig.de/). Collect *Pyrococcus* tRNA sequences. Make a "typical" and a consensus tRNA cloverleaf diagram. Then download the tRNA sequences in multi-FASTA alignment format. Go to Weblogo (http://weblogo.threeplusone.com/create.cgi) and create a sequence logo. Repeat for *Sulfolobus* and all archaeal tRNAs. Move images into PowerPoint to make figures. If you want to see something disappointing, repeat for bacterial tRNAs, in which patterns are badly degraded. That is evolution for you. Archaeal tRNAs are much more LUCA-like than bacterial tRNAs.

Chapter 25

The Three Domains of Life on Earth and Scientific Working Hypotheses

The purpose of this chapter is to generate a reasonable conceptual scientific working hypothesis for the three domains of life. Much of the material in this chapter is also found elsewhere in the book, but I thought it should also be consolidated here to focus the argument. If this book is worth reading, it is worth reading more than once, and key sections are worth revisiting. Some of the book content may bear repeating. This is, after all, a book about using iteration of sequences and motifs and sometimes pseudosymmetry to generate biological complexity.

No attempt is made in this chapter to be exhaustive in description of the last universal common (cellular) ancestor (LUCA) or the three domains: archaea, bacteria, and eukaryotes. This is a conceptual working model, not a complete description.

The purpose of a conceptual working model is that it is used to generate hypotheses for understanding and, if possible, for later testing. Experiments tend to be simple comparisons or tests, so one must think simply to design hypotheses and experiments. The concept of a useful and simple working model is at the core of scientific investigation.

LAST UNIVERSAL COMMON (CELLULAR) ANCESTOR

LUCA (Fig. 25.1) is described in many ways by different people. My strong feeling is that it is most important to try to grasp LUCA at the time of the great divergence to archaea and bacteria. If you argue, for instance, that LUCA lacked a cell membrane and relied on an RNA genome, to me, you are describing the pathway from the RNA-protein world to the DNA genome world. I find this description of LUCA to be unsatisfying, whether or not it is partly true. I consider LUCA as close to the first organism(s) with a DNA genome and the first organism(s) with an intact cell. LUCA is more a concept than an organism, anyway. Much can be inferred about LUCA, but little can be known with certainty, and philosophical modeling of LUCA may have limited use.

Evolution since Coding. http://dx.doi.org/10.1016/B978-0-12-813033-9.00025-1

GTFs:	RTase	Membranes?	TIM barrels $(\beta-\alpha)_8$
PIF $\geq$(HTH)$_4$	Topoisomerases		Rossmann folds $(\beta-\alpha)_8$
TBP	Gyrases		ATPases
RNAP (two DPBB-type)			GTPases
			Kinases
Promoters:			
BRE-TATA-BRE-TATA-BRE-TATA-BRE-TATA			Ribosome
			Cloverleaf tRNA
			20 amino acids

FIGURE 25.1 LUCA.

According to my description, LUCA has a reasonably intact and streamlined DNA genome, but a primitive DNA replication system that persists after the great divergence. LUCA is posited to replicate DNA by transcription followed by reverse transcription. Starting from a promoter DNA sequence, "transcription" is the synthesis of RNA from a DNA template. The LUCA promoter is posited to be a BRE-TATA-BRE-TATA-BRE-TATA-BRE-TATA repeat (at least four repeats) [BRE for transcription factor B (TFB) response element]; TATA for, i.e., TATAAAAG [a TATA box, which binds TATA-binding protein (TBP)]. TFB and TBP are archaeal general transcription factors (GTFs), factors that allow RNA polymerase to initiate transcription (RNA synthesis) from a promoter DNA sequence. TFB is a relic of the LUCA primordial initiation factor (PIF), which was at least a 4 helix-turn-helix (4 HTH) repeat. TFB has remaining 2 HTH domains, so archaeal TFB is posited to have lost at least 2 HTH repeats. It is possible that LUCA had other GTFs, such as an initiating helicase (to separate DNA strands), but this is not clearly known. The great divergence of archaea and bacteria occurred because of the differentiation of GTFs and promoters in the two systems. Modern replication systems are posited to have evolved separately and independently in archaea and bacteria after the great divergence, explaining why archaeal and bacterial replicative DNA polymerases are not homologous (genetically related).

Reverse transcriptase (RTase) is the enzyme that converts RNA to DNA. The first DNA genomes must have been generated using RTase, so this enzyme must have been present at LUCA and at the great divergence. After evolution of modern replication systems, RTase would disappear from archaeal and bacterial lineages. These cells likely could not tolerate two competing DNA replication systems.

LUCA had fairly modern metabolism. LUCA had TIM barrels, Rossmann folds, GTPases, ATPases, and kinases. At the time of the great divergence, LUCA had cloverleaf tRNA, probably 20 amino acids, and reasonably modern translation systems. It is not clear (to me) whether LUCA cell membranes were more similar to archaeal or bacterial cell membranes or a mixture of both.

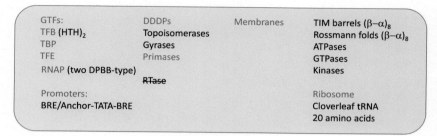

GTFs: DDDPs   Membranes   TIM barrels $(\beta-\alpha)_8$
TFB (HTH)$_2$ Topoisomerases   Rossmann folds $(\beta-\alpha)_8$
TBP   Gyrases   ATPases
TFE   Primases   GTPases
RNAP (two DPBB-type)   Kinases
                        ~~RTase~~

Promoters: Ribosome
BRE/Anchor-TATA-BRE   Cloverleaf tRNA
                                                                     20 amino acids

FIGURE 25.2   Archaea.

ARCHAEA

Archaea are posited to have diverged from bacteria because archaea utilize distinct GTFs and promoters (Fig. 25.2). At the time of the great divergence, the proto-archaea utilized transcription followed by reverse transcription to synthesize DNA. Therefore, promoters for RNA synthesis initiation were also replication origins, and GTFs, promoters, and RNA polymerase were even more important and central functions than they are today in archaea and bacteria. Because of the central importance of RNA synthesis in replication and transcription functions, divergence of GTFs, promoters, and RNA polymerase was sufficient to describe archaeal and bacterial divergence.

As GTFs, archaea utilize TBP (two TBP folds), TFB (2 HTH repeats), and TFE (two subunits with winged helix-turn-helix motifs). TFB is derived from the LUCA PIF. Bacteria committed to utilize sigma factors (4 HTH factors) derived from the LUCA PIF ($\geq$4 HTH).

After divergence, archaea evolved a replicative DNA polymerase (DDDP for DNA-dependent DNA polymerase) and shed RTase. This allows the streamlining and proper regulation of DNA replication. Archaea are similar to LUCA in their metabolism and translation systems. Archaea have distinct membrane systems compared to bacteria. It is not clear (till 2017) whether archaeal or bacterial membrane systems are more similar to LUCA membrane systems nor is it clear why archaeal and bacterial membrane systems diverged.

BACTERIA

Bacteria diverged from archaea after LUCA because of differences in GTFs (Fig. 25.3). Bacteria shed TBP, posited to be present at LUCA. Archaea retained TBP, which, in archaea, is posited to be not much changed since LUCA. To recognize a promoter DNA sequence, bacteria utilize sigma factors, which resemble the LUCA PIF in overall domain structure (4 HTH units remain in sigma 70), whereas archaeal TFB more closely resembles the LUCA PIF ($\geq$4 HTH) in TFB's remaining 2 HTH domain sequences and structures. Sigma factor HTH domains must fit within RNA polymerase so that RNA polymerase bound to

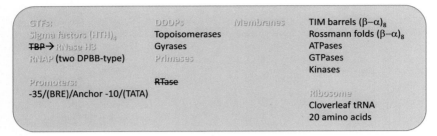

FIGURE 25.3 Bacteria.

sigma factor can recognize the promoter DNA sequence. Because sigma must make multiple interactions with RNA polymerase and promoter DNA, evolutionary pressures have altered and streamlined sigma HTH domains.

In the bacterial promoter, the most important DNA sequences are the Pribnow box (-10 region; i.e., $^{-12}$TATAATG^{-6}) and the anchor (-35 region; i.e., $^{-35}$TTGACA^{-30}). The Pribnow box is recognized by a modified HTH domain. I posit that the Pribnow box was initially a TATA box that bound TBP in the LUCA promoter but binds a modified HTH unit in the bacterial promoter. The -35 region in bacterial promoters binds the most C-terminal HTH$_4$ of the sigma factor. The -35 region corresponds to the BRE$_{anchor}$ or BRE$_{up}$ (TFB-recognition element upstream of TATA) of the archaeal promoter, which binds the most C-terminal HTH$_2$ of archaeal TFB.

I helped to sequence the first bacterial sigma factor gene in Richard Burgess' lab. I did not think I would live long enough to understand mechanisms of sigma factor function in transcription initiation, but combining evolution and structures, a clear story has emerged.

Bacteria have their own replicative DNA polymerase, posited to have evolved after divergence of archaea and bacteria. Bacteria have shed RTase, posited to have been present at LUCA, and RTase is posited to have been an essential part of the initial mechanism for DNA replication at LUCA.

Bacteria have membrane systems that are distinct from archaeal membrane systems. Bacteria have cloverleaf tRNA encoding 20 amino acids and ribosomes. Bacteria have typical metabolic pathways that probably mostly trace to the RNA-protein world before LUCA.

EUKARYOTES

In addition to other endosymbiotic fusions, eukaryotes are an unholy fusion of an α-proteobacteria bacterium and a Lokiarchaeota archaea (Fig. 25.4). The Lokiarchaeota archaea was occupied by a population of α-proteobacteria endosymbionts, which, over time, became the mitochondria. Group II introns from the α-proteobacteria attacked the Lokiarchaeota genome forming introns. The cell nucleus, the RNA polymerase II CTD, the vast CTD interactome, and the splicing apparatus for mRNA processing may have resulted as defenses against

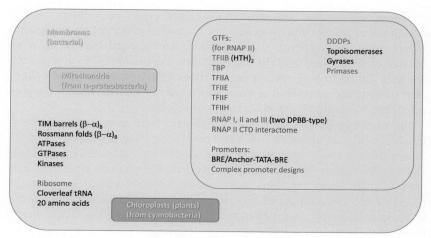

FIGURE 25.4   Eukaryotes.

the α-proteobacterial group II intron invasion, which also brought in a RTase/integrase, which subsequently had to be disabled. Complex cell signaling and epigenetics evolved in response to the endosymbiosis. Much of eukaryotic cell and organismal complexity was licensed as defenses against the archaeal–bacterial endosymbiotic fusion.

Plants became photosynthetic through endosymbiosis by cyanobacteria that became the chloroplast.

Eukaryotes are similar to bacteria in membrane composition. Much of eukaryotic metabolism is most related to bacterial systems. Transcription and replication systems are more archaeal in origin. Eukaryotes have cloverleaf tRNA, 20 amino acids, and ribosomes that are more archaeal than bacterial.

Of course, eukaryotic cells and eukaryotes are more complex than I describe here. Much of eukaryotic complexity, however, evolved initially as a defense against α-proteobacterial group II intron invasion of the Lokiarchaeota genome. The resulting innovations were repurposed to generate many features of the eukaryote cell, signaling, and organismal complexity.

DESCRIPTIONS AND PREDICTIONS

If this is a useful working model for genesis of life on earth, the model should provide clear descriptions and make potentially useful predictions. The model provides a coherent pathway for evolution of life on earth that is conceptual. The model is also simplifying. The model provides a surprisingly simple hypothesis for genesis of GTFs and promoters. In this way, the model is already useful. Please feel free to add to or edit the model if this makes it more useful to you.

The model predicts that iteration of sequences is a major mode for development of biological complexity. I would suggest that this prediction is justified by data, some of which are found in this book.

The model suggests that some of the largest events in evolution, such as divergence of archaea and bacteria, occur because of major changes in genome interpretation: replication and transcription. Archaea and bacteria diverged because of differences in GTFs and promoters. At LUCA, GTFs and promoters were central to both replication and transcription. In evolution of eukaryotes, many changes occurred in transcription and replication systems, allowing eukaryotes to develop into more complex organisms.

It is not known whether LUCA membranes were more archaeal or bacterial. I posit that alterations in membranes may have accompanied evolution of distinct DNA replication systems in archaea and bacteria. Looking at DNA replication systems and their interactions with the cell membrane may provide some insight into this major question about archaeal and bacterial evolution and differentiation. As I write (2017), I do not know whether this hypothesis is correct, but it occurred to me by designing the simple model (Figs. 25.1–25.4).

Chapter 26

LUCA

One of the mysteries of the transition from the RNA genome world to the DNA genome world is that archaea and bacteria, which must both be rooted in the last universal common ancestor (LUCA), lack homologous DNA polymerases.

To me, this is not a problem: this makes perfect sense.

To transition from an RNA genome world to a DNA genome world initially required mostly two functions: (1) a two double-Ψ–β-barrel-type RNA-template-dependent RNA polymerases (we are intimately acquainted with one: PDB 2J7O) (Fig. 17.3) and (2) a reverse transcriptase to convert RNA to DNA and then to wrap back and synthesize the complementary DNA strand.

The selective advantage to the DNA genome world is the greater stability of double-stranded DNA compared to single-stranded RNA. The advantages of DNA are obvious. For genome stability, what a difference a 2′-OH (present in RNA but missing in DNA) makes. The 2′-OH makes RNA much less chemically stable than DNA. Also, having a fully complementary strand makes DNA far less reactive than RNA.

RNA could not be replaced entirely by DNA because RNA is necessary for ribosomal RNA, transfer RNA, messenger RNA, ribozymes, protein synthesis by ribosomes, and many, many other functions. After DNA, the volatility of RNA was an advantage because, for one thing, RNA stability and/or turnover could be regulated as a means of gene regulation.

Once DNA copies of RNAs become possible, an evolutionary race commences to build the first streamlined, rapidly replicating, intact DNA genomes. To accomplish this evolutionary explosion, many functions become necessary. DNA topoisomerases become essential to underwind DNA. As Watson and Crick described, double-stranded, double-helical DNA is puzzlingly inert because the bases that carry genetic information are hidden on the inside of the helix. From the inception of DNA genes, to replicate, transcribe, and recombine DNA requires double-helix opening. This is partly the job of a DNA topoisomerase of the DNA gyrase type, which is partly a $(\beta$–$\alpha)_{4-5}$ repeat refolded from a larger $(\beta$–$\alpha)_n$ repeat. Such α/β repeat proteins that include TOPRIM domains are now familiar. DNA ATP-dependent helicases may be important to open up the DNA helix. Probably, relatively short stretches of DNA were circular, double-stranded, negatively supercoiled (underwound), and packaged in proteins that may also have assisted DNA opening.

Evolution since Coding. http://dx.doi.org/10.1016/B978-0-12-813033-9.00026-3

113

DNA ligation and recombination mechanisms are necessary to assemble large DNA genomes from parts. DNA ligases resemble RNA ligases, so it is clear where DNA ligases come from. Because I am not expert in DNA recombination, I leave this important issue for now. RecA DNA recombinases are α/β repeat proteins.

For our story, we consider LUCA as the first cellular organism and the first organism with a streamlined DNA genome. Other ideas about LUCA are possible, but what I most wish to explain is the divergence from LUCA to archaea and bacteria. At the time of the great divergence, LUCA was likely cellular and likely had an intact, circular DNA genome.

LUCA likely included DNA recombination mechanisms, sliding replication clamps, DNA ligases, reverse transcriptases, and DNA topoisomerases. It had ribosomes, ribosomal RNA, transfer RNA, messenger RNA, and ribozymes. It had glycolysis, a citric acid cycle, a glyoxylate cycle, diverse metabolism, ATPases, GTPases, and kinases.

A lingering mystery of the divergence of archaea and bacteria from LUCA is that archaea and bacteria have very different membrane lipids. As I write (2017), I do not have a simple explanation that I favor for this observation.

From the ages of microfossils of cellular organisms, LUCA existed on earth ~3.5 to 3.8 billion years ago. The RNA world and the RNA-protein world may have existed ~4.1 billion years ago. The earth is ~4.6 billion years old in a ~13.8 billion-year-old universe. For the current discussion, the exact dates do not matter much.

Chapter 27

General Transcription Factors and Promoters

So, how much might we know or surmise about the divergence of archaea and bacteria ~3.5–3.8 billion years ago?

My hypothesis is that divergence of archaea and bacteria was primarily driven by a fundamental difference in the evolution of general transcription factors, multi-subunit two double-Ψ–β-barrel-type RNA polymerases and promoters. According to the model I shall present, archaeal transcription more closely resembles the last universal common (cellular) ancestor (LUCA) transcription. Bacterial transcription diverged more radically to become more streamlined. In bacterial transcription, RNA polymerase, general transcription factors, and promoters are more powerfully coevolved and codependent than in archaea.

The model is shown in Figs. 27.1–27.3. Fig. 27.1 shows a model for the primordial LUCA general transcription factors and promoter. The promoter is the DNA sequence from which transcription (RNA synthesis) initiates. Iteration of an helix-turn-helix (HTH) unit generates a ≥4 HTH primordial initiation factor that binds to a BRE-like DNA sequence [BRE for transcription factor B (TFB)-recognition element]. TFB is a 2 HTH factor in archaea derived from a 4 HTH primordial initiation factor. At LUCA, TATA-binding protein (TBP; a two TBP fold repeat protein) is posited to be largely the same as archaeal TBP. The LUCA promoter is posited to be a linear alternating repeat of BRE and TATA sequences. Because these transcription factors are widely distributed in archaea and bacteria, LUCA is posited to have had dimeric HTH and winged helix-turn-helix (WHTH) factors. So, LUCA general transcription factors and promoters are generated from a small number of protein domains and interacting DNA sequences.

Fig. 27.2 shows a model for evolution of the archaeal general transcription factors and promoter from the posited LUCA system. To suppress inaccurate transcription starts, the archaeal promoter is simplified from the LUCA promoter. TFB is posited to lose ≥2 HTH motifs from the ≥4 HTH primordial initiation factor. Each TFB HTH binds a BRE on either sides (upstream and downstream) of the TATA box, which binds TBP. After >3.5 billion years, the two HTH repeats in archaeal TFB remain very recognizable from inspection of linear sequence. TFB is modified at the N-terminal end with a Zn finger. TFEα and TFEβ are WHTH factors that are recruited as general transcription factors. TFEα has a C-terminal Zn ribbon. Archaea have dimeric HTH and WHTH transcription factors.

Evolution since Coding. http://dx.doi.org/10.1016/B978-0-12-813033-9.00027-5

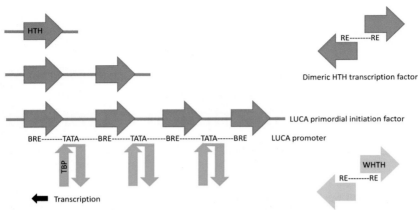

FIGURE 27.1 A model for evolution of the last universal common (cellular) ancestor (LUCA) general transcription factors and promoters. A LUCA primordial transcription initiation factor is posited to be a linear repeat of ≥4 helix-turn-helix (HTH) units (arranged N→C) that binds a BRE-related DNA sequence. TATA-binding protein (TBP) is a linear repeat of two TBP folds that binds to a TATA box (i.e., TATAAAAG) DNA sequence. LUCA is posited to have included dimeric HTH and winged helix-turn-helix (WHTH) factors. The direction of transcription is indicated (*small black arrow*). RE for response element (DNA to which a transcription factor binds).

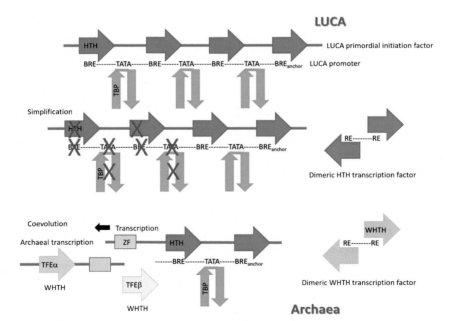

FIGURE 27.2 A model for evolution of archaeal transcription. Archaeal transcription is generated from the last universal common (cellular) ancestor (LUCA) transcription via simplification and coevolution. The primordial initiation factor is simplified to two helix-turn-helix (HTH) units in archaeal transcription factor B (TFB) binding to BREs. An N-terminal Zn finger (*yellow rectangle*) is added to TFB. TATA-binding protein (TBP) is preserved binding to a TATA box (i.e., TATAAAAG) DNA sequence. TFEα and TFEβ evolve from two different winged helix-turn-helix (WHTH) factors. Archaea include dimeric HTH and WHTH transcription factors.

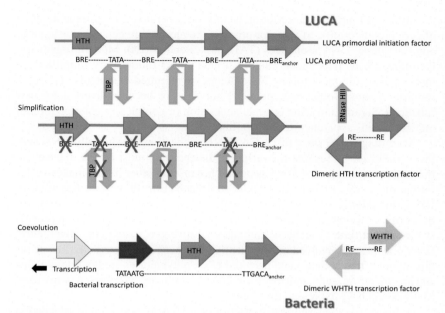

FIGURE 27.3   A model for evolution of bacterial transcription. Bacterial transcription is generated from the last universal common (cellular) ancestor (LUCA) transcription via simplification and coevolution. The ≥4 helix-turn-helix (HTH) primordial initiation factor becomes a 4 HTH bacterial σ factor. Sequence changes render the most N-terminal σ factor HTH divergent and the second HTH becomes specialized to recognize a Pribnow box (i.e., TATAATG) DNA sequence posited to be a relic of a TATA box (i.e., TATAAAAG).

Fig. 27.3 shows a model for evolution of the bacterial general transcription factors and a bacterial promoter from the proposed LUCA system. Bacteria are posited to have diverged from archaea because of the divergence of general transcription factors and promoters. It is important to note that at LUCA, RNA polymerase promoters doubled as replication origins, and this dual function persisted even after the great divergence to archaea and bacteria (see below). Four HTH bacterial σ factors are posited to be derived from a LUCA ≥4 HTH primordial initiation factor through degeneration of HTH repeats. In σ factors, HTH_1 is highly divergent and has essentially lost its mission as an HTH that binds DNA. HTH_2 has become specialized to open the promoter at the Pribnow box DNA (i.e., $^{-12}TATAATG^{-6}$), which is proposed to be derived from a TATA box (i.e., TATAAAAG). HTH_3 and HTH_4 of σ factors are very recognizable HTH units that interact with bacterial promoter DNA. Bacteria no longer utilize TBP or a TATA box (except for the Pribnow box, which is now recognized by σ HTH_2). Although bacteria lack TBP, bacterial RNase HIII is a protein with a single TBP fold, indicating that a bacterial ancestor (i.e., LUCA) may have included TBP, as I posit here. TBP is a repeat of two TBP folds.

Models for general transcription factors are generated using simple iteration of core protein motifs. Models for promoter sequences are generated using iteration of simple DNA sequences (TATA boxes and BREs). Modern organisms, therefore, are evolved via iteration of simple motifs followed by simplification through coevolutionary forces. New factors are recruited to systems via coevolutionary forces. Sometimes, parts are salvaged or become diverted for new functions. Significantly, the original generation of complexity via repetition of motifs remains apparent after >3.5 billion years.

According to the story of LUCA, DNA replication was initially RNA synthesis followed by reverse transcription. Indeed, this must be the mechanism for generation of the first DNA genomes because no other model makes sense. Furthermore, this remains the replication mechanism of retroviruses such as HIV1 (the AIDS virus), showing that this mechanism is feasible. I posit that transcription followed by reverse transcription remained the replication process at the ancient time of archaeal and bacterial divergence, and after divergence, DNA polymerases evolved independently in archaea and bacteria. Evidence for this model is that archaeal and bacterial DNA polymerases are not homologous and are, therefore, separately evolved. I do not claim this as completely my idea. Many others have expressed this view.

If this is so, then genome transcription (RNA synthesis) and replication (DNA synthesis) are tightly linked processes at LUCA and become less tightly linked soon after divergence of archaea and bacteria. At LUCA, because of the DNA replication mechanism, two double-Ψ–β-barrel-type RNA-polymerase promoters also function as replication origins. Notice that, as mentioned above, transcription followed by reverse transcription at LUCA is also the mechanism for replication of retroviral RNA/DNA today, showing the mechanism is reasonable. Understanding transcription by two double-Ψ–β-barrel-type DNA-template-dependent RNA polymerases at LUCA, therefore, describes the core functions of genome interpretation (transcription) and also genome replication.

Because life evolved from an RNA-protein world, RNA polymerases are central to evolution.

Because at LUCA the first DNA genomes were more heavily dependent on DNA-template-dependent RNA polymerases, than they subsequently became, two double-Ψ–β-barrel-type RNA polymerases, their protein cofactors and interacting DNA sequences are central to archaeal and bacterial divergence.

So, what was the constellation of general transcription factors present at LUCA and how might one know?

What is the structure of a primordial LUCA promoter?

How can one utilize knowledge of genetics, evolution, repeating motifs, and biology to recreate transcription/replication at LUCA? Is enough preserved in current DNA and protein sequences?

If you could approach a probable or a feasible answer, is this good enough?

So, how can one generate a working model for genesis of DNA genome-based life?

Let us start with a primordial promoter sequence.

We shall create the primordial promoter by iteration of existing promoter motifs: a TATAAAAG box and an associated BRE (TFB-recognition element) (Fig. 27.1). This strategy for primordial promoter design is very similar to the iteration evident in generation of $(\beta-\alpha)_n$ repeat proteins. Repeated sequences and protein motifs result from RNA and DNA replication duplication errors and/or gene fusions (i.e., RNA ligations).

Promoters are DNA sequences from which RNA polymerase initiates transcription.

At LUCA, primordial promoters were also replication origins because DNA replication was transcription followed by reverse transcription.

So, if a primordial promoter is a (TATA box-BRE)$_n$ repeat, this pushes TBP and TFB back into LUCA. TBP and TFB, proteins that bind to the TATA box and to the BRE (TFB-recognition element), therefore, are initially both transcription and replication factors: stewards of two major DNA genome functions.

Even from primary amino acid sequence, both archaeal TBP (a two TBP-fold repeat protein) and TFB (a 2 HTH motif repeat protein) are generated as obvious repeats (Fig. 27.2). Primordial TBP at LUCA is modeled as archaeal TBP of today. TBP sits on the TATA box and settles in the DNA minor groove, "like a saddle on a bucking bronco." TBP bends DNA at about a 80–90 degree angle. Primordial TFB is modeled as a ≥ 4 HTH repeat structure. Currently, in archaea, TFB is reduced to two HTH repeats. (If you wish to build the model for the LUCA primordial initiation factor with two HTH TFB repeats, you can, but it is awkward for modeling bacterial σ factors, some of which have at least 4–6 HTH units.) The TFB repeats are termed *cyclin-like* repeats, but, despite structural resemblance to cyclins (famed for their roles in regulating the cell division cycle in eukaryotes), functionally, in TFB, these are HTH DNA-binding motifs. HTH motifs bind in the major groove of DNA, in this case, at BREs. The archaeal TFB HTH_2 (called the "anchor" HTH) orients the binding of RNA polymerase and specifies the direction of transcription.

Generally, bacteria utilize σ factors, RNA polymerase, and a promoter DNA sequence to initiate RNA synthesis. The promoter DNA sequence includes a −10 Pribnow box (i.e., $^{-12}$TATAATG^{-6}) and a −35 region (i.e., $^{-35}$TTGACA^{-30}) (+1 is the transcription initiation site). Bacterial σ factors (4 HTH degenerate repeats) are posited to be derived from a regular ≥ 4 HTH repeat ancestor, such as the proposed LUCA primordial initiation factor (Fig. 27.3). Bacteria are posited to have lost TBP and its TATA boxes. A downstream TATA box is posited to have been converted to a Pribnow box for recognition by σ HTH_2. HTH_4 binds the −35 region of promoters. The σ factor HTH_4 is the anchor HTH and corresponds to archaeal TFB HTH_2 in promoter binding and function.

Because of the ease in generating genetic iterations, repeating sequences generate much of the most ancient and longest enduring complexity in the fabric of life.

So, the LUCA promoter is an alternating repeat of TATA boxes and BREs (Fig. 27.1). TBP binds TATA in the DNA minor groove. TBP bends the DNA 80–90 degrees at every TATA box. Each TFB HTH of ≥4 total binds a BRE in the DNA major groove. Between each HTH unit on TFB, a TBP binds a TATA box. The resulting model is a left-handed supertwist of DNA wrapped with TBP on the outside at bends and TFB HTH domains at BREs. You can generate the initial repeat as long as you like ($n \geq 4$).

The purpose of such an arrangement would be to separate the DNA strands for transcription and replication. At LUCA, for both processes, recruitment of RNA polymerase is the crucial step. A promoter is a DNA sequence to which RNA polymerase binds and from which RNA polymerase initiates transcription (DNA→RNA; RNA synthesis).

If you wish to use the wrapped DNA-TBP-TFB scaffold to recruit a DNA helicase, as an aid to open up the promoter/replication origin, you can model it that way. Negative supercoiling (underwinding) of a circular DNA will also help. This is accomplished using a DNA gyrase type DNA topoisomerase, which is built on an ancient α/β repeat fold (a TOPRIM domain).

According to my model, TBP, pre-TFB, TATA boxes, and BREs existed at LUCA. So HTH domains existed at LUCA.

Promoters initially doubled as both promoters for RNA synthesis and as replication origins for DNA synthesis.

According to my model, promoters were initially more complex and repetitive than they generally appear today. The promoters of today are posited to be simplified from a more iterative primordial promoter through coevolution of interacting factors and DNA. Despite this simplification, however, current archaeal and bacterial promoter DNA sequences are not much different from the primordial LUCA promoter/replication origin of ~3.5–3.8 billion years ago. So, remarkably, LUCA remains visible in modern promoter DNA sequences.

So, generating increased complexity requires sequence duplications: 2 repeats in TBP, 2 or 4 HTH repeats in primordial TFB (I favor 4 or >4), iteration of TATA-BRE (as many copies as you like, at least 4; Fig. 27.1). If you bring in a TOPRIM containing DNA gyrase to negatively supercoil DNA, this adds a ~$(\beta-\alpha)_4$ repeat that was cut from a longer $(\beta-\alpha)_n$ repeat.

If you do not now believe in dumb sequence iteration as a primary mechanism for generating biological complexity in ancient evolution, you are a very tough audience.

I would not claim that this model should be cast in stone. This model is, however, simplifying, feasible, and informative. The model provides a simple explanation for how archaeal and bacterial transcription evolved and how these processes relate. The model describes how archaea and bacteria diverged from LUCA, which, otherwise, is not adequately described.

An attraction of my model is that it roots TBP, TFB, HTH proteins, WHTH proteins, TATA boxes, and BREs in LUCA. This model also posits roles for

TBP, TFB, TATA boxes, and BREs in the packaging, interpreting, and replicating of the earliest DNA genomes, before and even after divergence of archaea and bacteria.

My model suggests that the major key to understand the divergence of archaea and bacteria from LUCA (the rooting of the tree of life) is to analyze two double-Ψ–β-barrel type RNA polymerases, general transcription factors (TBP, TFB), and promoters (TATA boxes and BREs). Because, at LUCA, replication (DNA synthesis) is so tightly coupled with transcription (RNA synthesis), differentiating RNA polymerases, their general transcription factors, and primordial promoters/replication origins are sufficient to drive the differentiation of the archaeal and bacterial domains. Differentiated DNA replication mechanisms (i.e., DNA polymerases) in archaea and bacteria probably came later after divergence.

A couple of years ago (i.e., 2013), I would have said that to generate this type of a model for ancient evolution of promoters and replication origins would be impossible.

Now, I consider this to be a simplifying and empowering (and possibly accurate) working model for genesis of DNA genome-based life. If the story I relate has flaws, I cannot identify them. I believe a primordial promoter and general transcription factor set could now be constructed in a laboratory to work for accurate initiation of RNA synthesis in a test tube. This model makes many, many predictions for DNA sequence analyses, bioinformatics analyses, and experiments.

The story of life on earth largely reduces to this: iteration of sequence motifs→interactome→coevolution→simplification (or not).

In case I have not stressed this point, in the Old Testament of gene regulation, we have generated halos (barrels) $((\beta-\alpha)_8$ TIM barrels and cradle-loop barrel folds), cradles (cradle-loop barrel folds: RIFT barrels, double-Ψ–β-barrels and swapped-hairpin barrels), and wings (WHTH motifs).

TRY THIS AT HOME

Using Chimera, align PDBs 5IY7 chain A, 5I2D chain D, and 5BYH chain D (Fig. 27.4). This aligns the largest RNA polymerase subunit for the three structures forcing other parts of the structures with less sequence similarity into alignment. Retain all nucleic acid chains and Mg. Remove all protein chains except for 5IY7 chain M (human TFIIB), 5I2D chain F (*Thermus thermophilus* σ70 or σ^A), and 5BYH chain M (*Escherichia coli* σ54). Inspect the structure for the colocalization of HTH units. TFIIB is a string of two HTH units with an N-terminal Zn finger. σ70 is a string of ~4 HTH units. σ54 is a string of ~6–7 HTH units (depending on how you count them). You have just (remarkably) demonstrated homology of human TFIIB and bacterial sigma factors, demonstrating the close kinship of all living things on earth. Because TFIIB is homologous to TFB of archaea, this alignment shows the relatedness of bacteria and

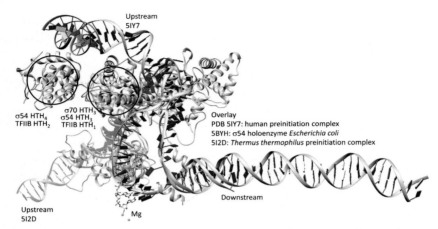

FIGURE 27.4   Overlay of PDBs 5IY7, 5I2D, and 5BYH. The human preinitiation complex containing TFIIB (5IY7) is beige. The bacterial preinitiation complex containing σ70 is blue (5I2D). The bacterial σ54 holoenzyme (5BYH) is pink. Homologous TFIIB HTH_1, σ70 HTH_3, and σ54 HTH_3 colocalize (*circle*) just upstream of the transcription bubble. Homologous TFIIB HTH_2 and σ54 HTH_4 also colocalize (*circle*). Homologous σ70 HTH_4 binds in another position determined by the trajectory of upstream DNA (occluded in the view shown).

archaea, describing how bacteria and archaea diverged from LUCA. You have also probably demonstrated that the primordial initiation factor at LUCA was probably a string of more than 6 HTH units, maybe ~$(HTH)_8$. This happy task of aligning sigma structures was kindly assigned to me by Martin Buck, Imperial College London, who works on σ54.

Chapter 28

Archaea

So, how do we get from the last universal common ancestor (LUCA) to archaea? Really, that is too easy.

Viewed as LEGO (Trademark) life, archaea looks a lot like LUCA (Compare Figs. 27.1 and 27.2).

The archaeal promoter includes a BRE_{up} (BRE upstream of TATA), a TATA box, a BRE_{down} (BRE downstream of TATA), and an initiator element (recognized as the transcription start by RNA polymerase). Essentially, the modern archaeal promoter can be generated from the LUCA primordial promoter through the coevolution of RNA polymerase, TBP, TFB, and promoters. TFE, with two WHTH domains and a Zn finger, was recruited to help support accurate initiation. Archaeal RNA polymerase is more ornate than single-subunit double-Ψ-β-barrel-type RNA-dependent RNA polymerases of the RNA-protein world. TBP is essentially unchanged. TFB is posited to have lost at least 2/4-cyclin-like repeats (HTH domains) and gained a Zn^{2+} finger near the N-terminal end. To support initiation, the TFB Zn finger and TFE may compensate for the loss of 2/4 HTH units. The promoter reduces to the modern archaeal promoter to optimize the accuracy of initiation by eliminating false, multiple RNA starts.

To this day, archaeal promoters are overall AT-rich, which is good, because AT base pairs are easier to open than GC base pairs, facilitating transcription. High AT content of archaeal promoters may relate to polymerization of the TATA box during ancient evolution, as I propose. The initiator element probably evolved to its current form through coevolution of RNA polymerase and promoters. TATA box repeats may have been eliminated over time via purifying (negative) selection to reduce false transcription starts.

Archaea had and have everything that LUCA and the RNA-protein world possessed in terms of ribosomes, ribosomal RNA, transfer RNA, messenger RNA, metabolism, ancient $(\beta-\alpha)_n$ repeat proteins, double-Ψ-β-barrels, RIFT barrels, diverse redox chemistry, ATP synthases, ATPases, GTPases, and kinases. Some ribozymes were sacrificed in the transitions.

Winged helix-turn-helix (WHTH) proteins are plentiful in archaea and bacteria. WHTH proteins probably existed at LUCA. TFE represents two separate lineages of WHTH factors.

Evolution since Coding. http://dx.doi.org/10.1016/B978-0-12-813033-9.00028-7

It is posited that ancient archaea may have possessed H3/H4 chromatin. Some archaea possess H3/H4 chromatin today.

This was a happier time for mesophilic archaea. For competition, there were no eukaryotes, which can be considered giant supercharged (mitochondria-driven) archaea, there were only bacteria as competitors.

Chapter 29

Bacteria

So, from our model for the last universal common ancestor (LUCA), how do we generate bacteria (Fig. 27.3)?

This is a story that resonates with me personally because I was part of a team that was the first to sequence DNA encoding a bacterial σ factor: *Escherichia coli* σ70. I did this with Richard Burgess, Carol Gross, and others at the University of Wisconsin, Madison. In those dark days, you had to sequence the DNA you struggled to clone. By modern standard, ours was a primitive operation.

Understanding bacteria requires an explanation of σ factors generated from ≥4 HTH pre-TFB/pre-σ primordial initiation factor. The number of HTH units in the primordial initiation factor is in question because σ70 appears to have 4 HTH units, but σ54, which is otherwise similar, appears to have 6–7 HTH units. Therefore, the primordial transcription factor may have had ≥6–7 HTH units.

The story of bacteria is a story of coevolution of multi-subunit RNA polymerases of the two double-Ψ–β-barrel types, σ factors, and promoters.

My son Sam and I told much of this story in 2014 in the journal *Transcription*. Aravind (NCBI) and coworkers contributed to this story.

At the heart of the tale is homology between the HTH domains $(HTH)_{1-4}$ of bacterial σ factors and archaeal TFB $(HTH)_{1-2}$. *Homology* indicates a common ancestry of proteins. Bacterial σ factors were derived from the posited ≥4 HTH pre-TFB/pre-σ primordial initiation factor. Bacterial σ factors were initially ≥4 HTH factors. The 4 HTH pattern remains clearly visible in the RpoA σ factor of *Thermus thermophilus*.

Since LUCA, the cyclin-like repeats (HTH domains) of archaeal TFB have not changed very much in amino acid sequence. By contrast, the 4 HTH repeats of bacterial σ factors have changed a great deal. *Degeneration* of the 4 HTH repeat pattern in σ (degeneration in sequence of the individual repeats) is explained by coevolution of RNA polymerase, the σ factor and the promoter.

Archaeal RNA polymerase cooperates with general transcription factors TBP, TFB, and TFE (Fig. 27.2). A promoter DNA sequence includes BRE_{up}, TATA box, BRE_{down}, and initiator element. TFE is a WHTH (winged helix-turn-helix) and Zn^{2+} finger factor. In terms of our ancient evolution model, archaea make perfect sense.

Evolution since Coding. http://dx.doi.org/10.1016/B978-0-12-813033-9.00029-9

In bacteria, the story is somewhat different, explaining divergence of bacteria and archaea. Bacterial transcription is mostly RNA polymerase, σ factors, and promoters. Because bacteria have fewer general transcription factors, according to the proposed chronology, TBP was jettisoned as a GTF in bacteria. Bacteria do, however, possess a relic of TBP, because bacterial RNase HIII is built on a very recognizable TBP fold. In bacteria, after establishing the dominance of σ factors to control transcription initiation, TBP evolved into or persisted as an enzyme to degrade RNA from RNA/DNA hybrids (bacterial RNase HIII). This may be an example of the genetic salvage of a perfectly good TBP protein fold after (or before) its primary initial transcriptional function was remaindered in bacteria.

It is not clear how to think about archaeal TFE. Both archaea and bacteria have WHTH factors and Zn^{2+} fingers. Multi-subunit RNA polymerases of the two double-Ψ-β-barrel types include multiple Zn^{2+} fingers. Zn^{2+} fingers are ancient protein motifs that must have been present at LUCA and perhaps in the emergence from the RNA-protein world.

In becoming strongly reliant on σ factors to recognize promoters, bacterial RNA polymerase became or stayed simpler than more ornate archaeal RNA polymerase. In bacteria, powerful coevolutionary forces molded RNA polymerase, σ factors, and promoters, resulting in specialization and simplification. Through coevolutionary forces, in order to recognize promoters, σ factors became most reliant on HTH_2 and HTH_4. HTH_2 is a specialized HTH domain that opens the −10 region Pribnow box of the bacterial promoter ($^{-12}$TATAATG^{-6}). HTH motifs are actually helix (H1) turn (T1) helix (H2) turn (T2) helix (H3) motifs. HTH_2 of σ factors is most degenerate in its helix 2 (H2), which is elongated and bent. H3 of σ HTH_2 has become very aggressive at attacking the nontemplate DNA strand and flipping out a specific base to induce promoter opening. The Pribnow box has the consensus sequence $^{-12}$TATAATG^{-6}. σ HTH_2 attacks the Pribnow box thus: $^{-11}$A is first flipped out followed by the flipping out of $^{-7}$T and the unzipping of DNA to +1 for transcription initiation. In contrast to HTH_2, σ factors HTH_3 and HTH_4 are typical HTH motifs. HTH_3 binds to the extended −10 region of promoters and corresponds to HTH_1 of archaeal TFB, which binds BRE_{down}. σ HTH_4 binds the −35 region ($^{-35}$TTGACA^{-30} consensus) of bacterial promoters. σ HTH_4 corresponds to HTH_2 of archaeal TFB, which binds BRE_{up}.

In bacterial σ factors, HTH_1 is mostly vestigial and is now mostly cooperative with the highly specialized HTH_2. Some σ factors have lost segments of HTH_1.

For promoter recognition, HTH_3 is generally less important than HTH_2 and HTH_4. HTH_3 is disappearing from some alternate σ factors that recognize alternate promoter DNA sequences.

The Pribnow box $^{-12}$TATAATG^{-6} that binds σ HTH_2 resembles the archaeal TATA box TATAAAAG that binds TBP. If the primordial promoter is posited to be a repeat of TATA boxes, it is not difficult to imagine how an archaeal promoter TATA box and a bacterial promoter Pribnow box can be derived from a primordial promoter. I posit that the Pribnow box of bacterial promoters was derived from a downstream TATA box in the LUCA promoter.

The evolutionary model, based on archaeal TFB-bacterial σ factor homology, therefore, provides a simplifying view of general transcription factors, promoters and RNA polymerases in archaea and bacteria.

This model can be expanded to describe many alternate promoters and alternate σ factors in bacteria.

It goes without saying that bacteria had and have essentially everything that LUCA and the RNA-protein world possessed in terms of ribosomes, ribosomal RNA, transfer RNA, messenger RNA, metabolism, ancient $(\beta-\alpha)_n$ repeat proteins, double-Ψ-β-barrels, RIFT barrels, swapped-hairpin barrels, diverse redox chemistry, ATP synthases, ATPases, GTPases, and kinases. Many ribozymes were remaindered in the evolution of LUCA, archaea, and bacteria.

To streamline genome replication, bacterial DNA polymerase replication systems were likely evolved separately after the great divergence of bacteria and archaea, explaining nonhomology of bacterial and archaeal DNA polymerases.

In short, the archaeal strategy was to retain TBP, simplify TFB, recruit TFE, and simplify the promoter to more precisely specify initiation sites.

The bacterial strategy was to remainder TBP, couple σ factors more tightly to RNA polymerase and to simplify and streamline the promoter. In bacteria, the σ factor is intimately entwined with RNA polymerase and σ HTH_2 and HTH_4 are generally most important for recognition of the promoter.

There are many additional layers to this story, but they are told elsewhere. They are hugely interesting but peripheral to the central narrative.

Chapter 30

Methane and Oxygen

As in the present day, in some ancient times, the biosphere controlled the atmosphere and hydrosphere (the oceans). In 2017, a single political party in one country deliberately changes the climate of earth (the atmosphere and hydrosphere) for money. In other ancient times, the geosphere (massive volcanic activity) controlled the atmosphere and hydrosphere.

Before the birth of eukaryotes (~1.6–2.1 billion years ago), methane was a significant atmospheric component. Compared to CO_2, methane is a potent greenhouse gas that warmed the planet. Methane was produced by anaerobic bacterial methanogens. Then as now, the biosphere controlled the atmosphere.

Cyanobacteria generate oxygen and also developed the capacity for photosynthesis about 2.4 billion years ago.

Initially, O_2 was held in the ground as rust (Fe_2O_3 and Fe_3O_2). Finally, the elemental Fe was depleted, and O_2 could no longer be sequestered as rust; therefore, free O_2 gas started to accumulate in the atmosphere. About 200 million years of cyanobacterial photosynthesis was required for O_2 to react with all of the available elemental Fe of earth. This is considered the Great Oxygenation Event or GOE, and photosynthetic cyanobacteria are responsible. The GOE is described as the *oxygen catastrophe*. For most anaerobic organisms, oxygen is catastrophic: a deadly toxin. Because archaea often lack oxidative phosphorylation, many archaea are anaerobes. Many bacteria are anaerobes or can switch between an anaerobic and aerobic existence. Cyanobacteria generated O_2. Then as now, the biosphere controlled the atmosphere.

And:

$$CH_4 + 2O_2 \leftrightarrow CO_2 + 2H_2O$$

But CH_4 is a more potent greenhouse gas than CO_2; therefore cyanobacteria and photosynthesis decreased atmospheric CH_4 and replaced it with CO_2 and O_2. Conversion of CH_4 to CO_2 caused the planet to cool.

In those days, the sun was dimmer than today, so trading O_2 and CO_2 for CH_4 in the atmosphere had the severe effect of global cooling.

The result was snowball earth (the Huronian Glaciation; ~2.1–2.3 billion years ago). The earth was shrouded in snow and ice. Snowball earth lasted ~200–300 million years.

Evolution since Coding. http://dx.doi.org/10.1016/B978-0-12-813033-9.00030-5

The biosphere controlled the atmosphere. A known change in the biosphere, the advent of photosynthesis by cyanobacteria, flipped the gaseous components of earth's atmosphere, with a radical change in earth's climate and temperature.

Section II

The New Testament of Gene Regulation

Chapter 31

LECA

LECA is an amazing story that continues to grow and improve (2017) (Fig. 31.1).

LECA is a story of multiple endosymbioses, unholy mixing of genetic material, remarkable genetic violence, genetic revenge, and surprising drama.

Eugene Koonin (NCBI) and Aravind (NCBI) were the inspirations for my version of the story of LECA. The recent discoveries of Lokiarchaeota (2015) and related Asgard archaea (2017) contribute to this story. A recent study in *Nature* by Pittis and Gabaldon (2016) enriches the model, but makes it more confusing. For instance, both Lokiarchaeota and α-proteobacteria appear to be late entries into the strange and surprisingly complex chimeric fusion that became eukaryotes. The story appears to be of serial endosymbioses in which archaea engulfed archaea and archaea engulfed bacteria, ultimately leading to a complex eukaryotic cell with a strange patchwork of genomic influences and multiple cellular membranes and organelles. Remarkably, in the late step, a resident population of α-proteobacteria morphed into mitochondria to engender the large eukaryotic cell with its unprecedented energetics. Genetic fusion of α-proteobacteria and Lokiarchaeota archaea unleashed group II intron invasion of the archaeal genome, leading to eukaryotic gene introns and splicing.

LUCA is the last universal cellular common ancestor of archaea, bacteria, and eukaryotes (~3.5–3.8 billion years ago).

LECA is the last eukaryotic common ancestor (~1.8–2.2 billion years ago), and FECA is the first eukaryotic common ancestor. The tangled pathway from FECA→LECA is not yet completely described.

So, where do eukaryotes come from?

Based on the stem lengths of phylogenetic trees mostly drawn from protein system sequences, it appears that eukaryotes evolved over a rough minimum of ~4 stages. In an early stage, an actinobacteria (bacteria) may have been engulfed by a Crenarchaeota (archaea). The Actinobacteria may have become the nucleolus of a future eukaryotic cell. In a following phase, the cell nucleus was acquired by possible engulfment of a Euryarchaeota (archaea), which may have left behind the nucleus as a relic of the endosymbiosis. A subsequent multistep evolutionary phase resulted in acquisition of the Golgi apparatus and the endoplasmic reticulum. Probably two endosymbioses occurred with engulfment of two separate non-α-proteobacteria bacteria, for which the family identities are not so far clearly reported (2017). Perhaps, each symbiosis donated its bacterial

Evolution since Coding. http://dx.doi.org/10.1016/B978-0-12-813033-9.00031-7

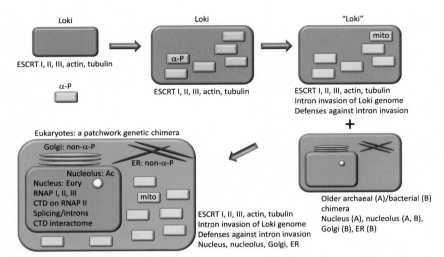

FIGURE 31.1 A possible model for the late stage of LECA evolution. Lokiarchaeota engulfed α-proteobacteria (α-P) to form the mitochondria (mito). The Lokiarchaeota/α-proteobacteria endosymbiont may have been rescued by endosymbiotic fusion to an older, complex archaeal/bacterial chimera that provided the nucleus, nucleolus, Golgi, and endoplasmic reticulum (ER). *Ac*, Actinobacteria; *Eury*, Euryarchaeota. The cell nucleus is a partial protection against α-proteobacterial group II intron invasion.

cell-membrane system to its archaeal host to form the Golgi and endoplasmic reticulum. In a final stage, Lokiarchaeota (archaea) and α-proteobacteria were engulfed, possibly as a preformed endosymbiont pair (Fig. 31.1). Without chimeric fusion to a cell that includes a nucleus, the Lokiarchaeota/α-proteobacteria endosymbiont might not have survived. With genes that contain introns, a nucleus becomes necessary to insulate premessenger RNA splicing in the nucleus from messenger RNA translation on the ribosome in the cytoplasm. The α-proteobacteria became the mitochondria. According to this ornate model, eukaryotes are a strange genetic chimera of many archaea and bacteria. Whenever very large packets of genetic material are transferred and a membrane system or organelle remains as a relic, consider endosymbiosis the likeliest gene transfer mechanism. Other mechanisms of horizontal gene transfer occur but often cannot transfer such robust packets of related and codependent genes and cannot transfer membrane systems. Thaumarchaeota and Korarchaeota are two other archaeal families that apparently transferred large packets of genes and may have resided as endosymbionts. Chlamydia and δ-proteobacteria are other bacterial families that appear to have donated large gene packets, possibly as endosymbionts during the late stage of eukaryogenesis.

According to this model, ≥6 endosymbioses (and possibly many more) may have led to the first eukaryotes. Strangely, mitochondria and Lokiarchaeota were both acquired late, inspiring a model in which an older complex chimeric cell with multiple organelle and membrane systems merges with a newer dimeric chimeric fusion (Fig. 31.1). Lokiarchaeota is capable of engulfing an endosymbiont, but many modern archaea have lost this capability, which they may, however, previously have possessed. Based on the model above, to form eukaryotes, multiple ancient archaea appear to have been capable of endosymbiotic engulfment of a bacteria, another archaea or an archaea–bacteria endosymbiont. As an analogy, consider Russian nesting dolls and *Matryoshka* endosymbiosis (John Stiller, East Carolina University) in which cells engulf others and are subsequently engulfed by others.

Although the story related above is much more complex, in the later stage, eukaryotes appear to be an unholy fusion of a Lokiarchaeota and an α-proteobacterium.

Before telling the story of the late fusion, I must relate the amazing discovery of Lokiarchaeota, reported in 2015. As you may know or might have guessed, Lokiarchaeota were named after the Norse god Loki, the treacherous trickster god. More specifically, Lokiarchaeota were named after the Loki Traps Deep Sea Hydrothermal Vents, near which multiple species of Lokiarchaeota were identified by shotgun DNA sequencing of organisms from undersea sediment cores. Prior to 2015, many thought that eukaryotes must be generated via endosymbiosis of an archaea (to be identified) and an α-proteobacterium (that became the mitochondria), but most known modern archaea do not have the apparatus necessary to engulf a bacterial endosymbiont (Fig. 31.2).

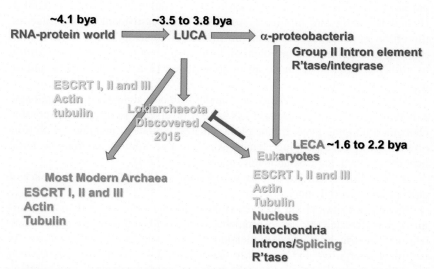

FIGURE 31.2   Engulfment of a bacterial endosymbiont by a Lokiarchaeota requires ESCRT systems, tubulin, and actin. These genes are lost in most modern archaea.

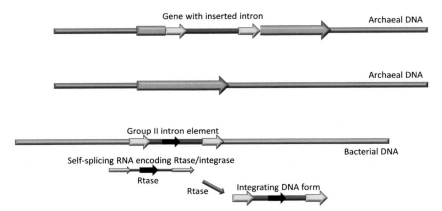

FIGURE 31.3 Invasion of the archaeal genome by bacterial group II introns.

Most modern archaea lack ESCRT (endosomal sorting complex required for transport) I, II, and III systems and also actin and tubulin. These are the systems required for one cell to engulf another to engage in endosymbiosis or phagocytosis (phagocytosis is one cell engulfing and "eating" another). Some other modern archaea bear remnants of some of these genes, possibly indicating that other parts of the endosymbiosis machinery were lost during evolution of most archaea.

Lokiarchaeota, however, include the ESCRT I, II, and III systems, actin and tubulin, making Lokiarchaeota much more similar to eukaryotes compared to many other archaea. Eukaryotes, of course, also utilize ESCRT I, II, and III, actin and tubulin systems for cell engulfment. Lokiarchaeota, therefore, appears to be one of the formerly "missing links" in evolution relating archaea and eukaryotes, and eukaryotes appear to be, in addition to multiple older influences, an unholy fusion of a Lokiarchaeota and an α-proteobacterium.

The late stage story relates how endosymbiosis unleashed unimaginable genetic violence by a semiselfish DNA/RNA/reverse transcriptase transposable element, a jumping gene. In one of the greatest clashes of genomes of all time on planet earth, the jumping element was activated from the α-proteobacterial genome to attack the Lokiarchaeota genome (Fig. 31.3).

It has long been known that the mitochondria of eukaryotes are the relics of an α-proteobacterial endosymbiont.

An ancient mesophilic archaea (Lokiarchaeota; now pushed to the margins) was invaded by an α-proteobacterial endosymbiont. Imagine many α-proteobacteria living and replicating within their Lokiarchaeota host. As α-proteobacteria live and die within the archaea, DNA is transferred between the two genomes. But the α-proteobacteria has a defense against foreign DNA: its group II intron mobile DNA element. Bacteria that include a group II intron element prevent it from hopping. Exposure to foreign archaeal DNA, however, allowed the group II intron element to hop, and hop it did.

There was a cataclysmic invasion of α-proteobacterial group II introns throughout the Lokiarchaeota DNA.

Group II intron elements are somewhat similar to retroviruses or transposons: jumping genes. Group II introns form integrating DNA elements that can jump into an unprotected DNA genome. They form self-splicing RNA elements. They encode a reverse transcriptase that can copy genetic material from RNA into double-stranded DNA, which can integrate again and again into an unprotected DNA genome.

The archaeal genome became littered with inserted group II introns. How the damaged Lokiarchaeota–α-proteobacterial fusion survived is somewhat difficult to imagine, but, apparently, at least one such experiment did survive. Of course, the process of archaeal–bacterial fusion could have been replicated many, many times with many less successful or very different outcomes, such as massive horizontal gene transfers. Engulfment of the Lokiarchaeota–α-proteobacterial endosymbiont by an older chimera with a nucleus may have saved the newer endosymbiont (Fig. 31.1).

The fused Lokiarchaeota–α-proteobacterial hybrid later faced multiple challenges. The archaeal genome needed to suppress the promiscuous hopping of group II intron elements, and suppression requires disarming the group II intron reverse transcriptase. To survive, the Lokiarchaeota genome needed to block intron insertions and command the splicing of introns. Therefore, the reverse transcriptase of the group II intron element needed to be inactivated. Furthermore, self-splicing of RNA introns needed to be replaced through the evolution of a novel machinery, which became the RNA polymerase II messenger RNA splicing apparatus. Also essential was the need to sequester the RNA splicing apparatus from the ribosome to avoid clashes between splicing and translation of messenger RNA on ribosomes. Thus, there was a powerful selective pressure to maintain the cell nucleus, which may already have been present from a previous endosymbiosis. The nucleus was needed as a barrier to separate splicing in the cell nucleus from translation of messenger RNA in the cytoplasm because translation of intron sequence is wasteful, nonsensical, and could be lethal. New eukaryote mechanisms needed to evolve to support the novel cell architectures, regulated splicing, and enhanced energetics.

Our primitive eukaryote now had a mature cell structure with a nucleus, nucleolis, endoplasmic reticulum, Golgi, and mitochondria. Much of the α-proteobacterial DNA was lost. Some was transferred to the archaeal genome. Some remains as mitochondrial DNA. Because the mitochondria enables aerobic respiration (oxidative phosphorylation), the chimeric fusion had capabilities to support cell energetics missing in the anaerobic mesophilic host. The fusion had more complex cell and genome structure.

Because of the massive Lokiarchaeota–α-proteobacteria genome fusion and the other chimeric genomic influences, evolutionary pressures on eukaryotes became very different from those on archaea and bacteria. The eukaryotic strategy was now to develop more complex cell compartmentalization, more sophisticated cellular communication mechanisms, more varied and potent energy metabolism, and more complex genomes, allowing for more complex and opportunistic lifestyles. Because of the mitochondria and enhanced energy generation, eukaryotic cells could be larger than archaeal and bacterial cells. Archaea and bacteria, by contrast to eukaryotes, were largely committed to rapid and efficient replication and mostly adhered to single-celled and small-celled lifestyles.

So, in competition for new and previously occupied niches, cell and genomic complexity and vastly enhanced energetics became potent strategies that set eukaryotes apart from archaea and bacteria.

Eukaryotes took over many environments previously filled by mesophilic archaea capable of endosymbiosis. Competition with eukaryotes caused archaea to shed ESCRT, actin, and tubulin systems for engulfing endosymbionts. Living as only slightly more complex prokaryotic cells, chimeric archaea could no longer compete with eukaryotes to occupy mesophilic niches.

Chapter 32

Eukaryotic Multi-Subunit RNA Polymerases, General Transcription Factors, and the CTD

My failed career as a biochemist/molecular biologist/academic has mostly centered on multi-subunit RNA polymerases of the two double-Ψ–β-barrel type (Chapters 18 and 19) and their general transcription factors (GTFs) (Fig. 32.1). I have worked on bacterial σ factors, eukaryotic TFIIF (transcription factor IIF; a general transcription factor), and bacterial and eukaryotic (i.e., yeast and human) RNA polymerases. This puts my career and interests on the superhighway of evolution from the RNA-protein world forward because the major events in evolution track two double-Ψ–β-barrel-type RNA polymerases and their interacting factors. From reading the work of Aravind (NCBI) and Koonin (NCBI) and others, I discovered a parallel universe of ancient evolution that fits closely with many partly formed ideas about transcriptional regulation, mechanisms, kinetics, and dynamics. Remarkably and unexpectedly, the integration of RNA synthesis systems and ancient evolution provides a simple, reasonably complete, and believable story of genesis.

With recently gained insights into ancient evolution, I begin to understand how to think about transcription mechanisms and their interactions within a larger cell, and the story makes frightening sense. New insights are simplifying and predictive, which gives me hope that these ideas might be of value. Many ideas in science appear to generate ever additional complexity with limited insight. By contrast, the story of genesis brings surprising clarity.

Because successful innovations survive, evolution gives retrospective purpose to biology. So, what can evolution tell us about two double-Ψ–β-barrel-type multi-subunit RNA polymerases and their GTFs, after unlikely but surprisingly successful fusion of a Lokiarchaeota (or a near relative) and a population of invading endosymbiotic α-proteobacteria?

First of all, there is much to explain. Why, at LECA, did eukaryotes differentiate functions into RNA polymerases I, II, and III (Fig. 19.4)? Why did the carboxy terminal domain (CTD) append to RNA polymerase II (Fig. 19.5)?

Evolution since Coding. http://dx.doi.org/10.1016/B978-0-12-813033-9.00032-9

139

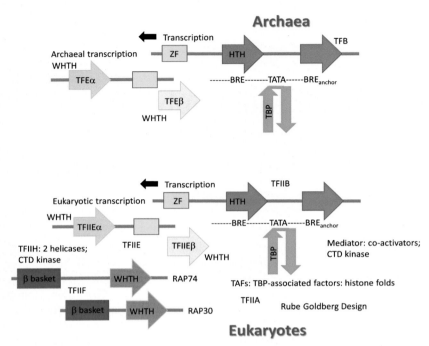

FIGURE 32.1 General transcription factors for eukaryotic RNA polymerase II.

How and why did a huge CTD interactome develop around the CTD? How did chromatin develop with its complex interactome and modification system (some of these are epigenetic systems)? How and why did ornate eukaryotic cell signaling develop? Why do the strange and extensive constellation of GTFs surrounding RNA polymerase II resemble a hare-brained Rube Goldberg device of endless add-ons and compensations (Fig. 32.1)? Is this the result of intelligent design or evolution?

I shall try to answer some of these questions.

In 2014, I addressed many of these issues in a review I wrote. If I wrote it today, I think I might change a few things, but not many: live and learn. I am more expert now in ancient evolution.

From the inception of eukaryotes, three RNA polymerases were present (I, II, and III) and a CTD heptad repeat was an appendage to RNA polymerase II (Figs. 19.4 and 19.5). In his 2014 PNAS paper, John Stiller (East Carolina University) convinced me this must be so. The CTD and the associated RNA splicing apparatus evolved as a defense against α-proteobacterial group II intron invasion of the Lokiarchaeota genome.

With a larger eukaryotic genome and greater tendency toward complexity, generation and maintenance of three RNA polymerases was easier for eukaryotes than for archaea and bacteria, which tend to streamline their genomes for rapid replication.

In eukaryotes, separating RNA polymerase functions among three polymerases allowed increased specialization and innovation in messenger RNA synthesis by RNA polymerase II. Complex cell structures, alternate lifecycles, and eventually eukaryote multicellularity require more nuanced gene regulation by RNA polymerase II.

Of course, with three RNA polymerases in eukaryotes, RNA polymerase I (ribosomal RNA) and RNA polymerase III (transfer RNA, 5S RNA, and some small regulatory RNAs) become uncoupled from RNA polymerase II (messenger RNA and some small regulatory RNAs) and can assume regulatory patterns of their own.

The CTD of RNA polymerase II initially evolved to support cotranscriptional RNA splicing (removal) of introns (noncoding RNA in the midst of the message that would make nonsense of the messenger RNA if it were translated on a ribosome) (Figs. 19.5 and 32.2). The CTD consensus repeat sequence $(YSPTSPS)_n$ ($n = 26$ repeats in yeast, 52 repeats in humans) then became a scaffold for evolutionary innovation related to many RNA polymerase II functions. Many CTD functions involve signaling: phosphorylation, acetylation, methylation, and ubiquitinylation. Modification of the CTD and cycling of accessory factors tracks the position of RNA polymerase II through a gene from initiation from a promoter to promoter escape to elongation (i.e., through a gene) to termination and recycling (Fig. 32.2). The CTD interactome communicates to chromatin to allow RNA polymerase II to pass through the gene and then to reestablish chromatin contacts to DNA after passage of RNA polymerase II. The CTD interactome functions in messenger RNA modification and

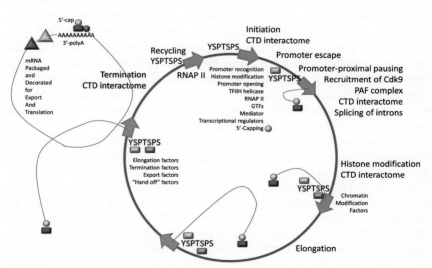

FIGURE 32.2 The RNA polymerase II (RNAP II) transcription cycle. RNAP II is depicted as a *fat blue arrow*. mRNA is depicted as a *thin blue line*. Red letters indicate phosphorylated YSPTSPS serines. Factors cycle on and off the carboxy terminal domain (CTD), depending on its modification state (i.e., serine 2 and 5 phosphorylation state) and the position of RNAP II in the cycle. This model is a vast oversimplification.

processing: capping, splicing, and polyadenylation. RNA synthesis termination factors that cleave the RNA at its 3'-end and dissociate RNA polymerase II from the DNA template associate with the CTD. The messenger RNA is bound by handoff factors, i.e., released from the CTD to the messenger RNA. These factors transport the messenger RNA through the nuclear pore to the cytoplasm and license the messenger RNA for translation on a ribosome.

The CTD on RNA polymerase II only makes sense as a scaffold for evolutionary innovation. The CTD is a story of generation of novel biological complexity via iteration of a simple protein motif. The CTD is a story of consistency (consensus 1-YSPTSPS-7 repeats; tyrosine–serine–proline–threonine–serine–proline–serine) and degeneracy (nonconsensus repeats). Consensus repeats generally interact with multiple factors in the vast CTD network interactome to help with RNA polymerase II–related functions. Because these repeats support diverse interactions, consensus repeats tend to be conserved and essential. Degenerate repeats may take on specialized functions. Particularly in complex organisms, nonconsensus repeats may assume specific developmental roles and factor interactions.

The CTD is a poster child for eukaryotic complexity. A scaffold for protein interaction is built by dumb iteration of a simple protein motif (1-YSPTSPS-7), and then other systems evolve to utilize that scaffold. Ultimately, the scaffold supports a host of interdependent functions and transactions. Chromatin, which packages DNA, can be viewed as a similar scaffold for evolution of interacting systems (epigenetics).

The extensive CTD interactome is linked to RNA polymerase II. This network interacts with the complex chromatin interactome to regulate DNA accessibility. Communication among these extensive interactomes coevolved with other signaling systems that regulate cell cycle and other processes. Evolution of these uniquely eukaryotic systems, therefore, supported complex signaling and communication to support more complex eukaryotic cells and multicellular organisms compared to archaea and bacteria, which rely on much simpler signaling networks. Coevolution of complex systems and communications licensed and drove the increased complexity of eukaryotes.

Consistent with the proposed model for Lokiarchaeota-α-proteobacterial fusion at LECA, eukaryotic RNA polymerases I, II, and III resemble archaeal RNA polymerase in subunit structure and amino acid sequence (Fig. 19.4). Eukaryotic GTFs for RNA synthesis also resemble the archaeal system rather than the bacterial system (compare Figs. 27.2, 27.3, and 32.1). Similar to archaea, eukaryotic RNA polymerase II utilizes TBP, TFIIB (related to archaeal TFB), and TFIIE (related to archaeal TFEα/β). In addition, RNA polymerase II utilizes TFIIF, TFIIA, and TFIIH. With Jack Greenblatt, Mary Sopta, and others, I helped with the cloning of human TFIIF subunit encoding genes. Similar to archaeal TFE and eukaryotic TFIIE, TFIIF is another WHTH (winged helix–turn–helix) factor. Suffice it to say that these are other examples of the higher complexity of eukaryotes (via add-ons) relative to archaea and bacteria. Sadly,

TFIIF, which I spent most of my career studying, appears to be another eukaryotic Rube Goldberg add-on.

In eukaryotes, TBP interacts with many TBP-associating factors. A large coactivator/corepressor of transcription known as mediator is a further eukaryotic embellishment. Without going through every polypeptide known to be involved, the overall RNA polymerase II functional network resembles a Rube Goldberg contraption of interacting activation, repression, antirepression, and anti-antirepression mechanisms. This system does not smack of intelligent design. This system appears to be cobbled together with molecular chewing gum and bailing wire: the stuff of dumb evolution. Looking at eukaryotic transcription from a human design point of view, it is amazing we survive. Although there is vast complexity, there appears to be no cleverness in our conception or construction.

The CTD, extensive CTD interactome, chromatin interactome, and the GTFs for RNA polymerase II are wonderful and telling examples of eukaryotic evolution and eukaryotic complexity. This is the opportunistic genetic junk of which we eukaryotes are made. Complex transcription systems license the complexity of eukaryotic cells and body plans. Much of complex eukaryotic signaling evolved around chromatin and transcription systems.

Chapter 33

Patchwork Eukaryotic Phylogenomics

Overall, the eukaryotic genome is a strange quilt of genetic influences with an essential core archaeal component and an even larger bacterial component. Many unique eukaryotic functions are also present. In terms of archaeal contributions, eukaryotes appear to be a genetic patchwork derived from multiple extant archaea. Complexity may largely be explained by a model of multiple nested endosymbiotic fusions (Fig. 31.1). Thus, Crenarchaeota appear to have contributed protein systems to the eukaryotic cell nucleolus. Euryarchaeota appear to have donated components to the nucleus. Some chromosomal proteins appear to have a Korarchaeota origin. Lokiarchaeota appears to have donated cell engulfment functions, but these functions must have been more broadly present in many archaea to have supported the older posited endosymbiotic fusions. Replication, transcription, and translation systems are archaeal in origin. In terms of bacterial contributions, major contributors appear to include Actinobacteria (nucleolus), weakly defined (2017) non-α-proteobacteria (at least two families: Golgi and endoplasmic reticulum), δ-proteobacteria, Clamydia, and α-proteobacteria (mitochondria). Given the complexity of the model, expect additions in the future. Eukaryotic membrane systems and metabolic systems are largely bacterial. Endosymbiosis appears to be a dominant mechanism to explain eukaryogenesis, because endosymbiosis transfers large packets of related genes and membrane systems (i.e., nucleolus, nucleus, Golgi, endoplasmic reticulum, mitochondria). Genetic fusions and resulting clashes of genomes and systems drove many eukaryotic innovations, for instance, to cope with the α-proteobacterial group II intron invasion of the Lokiarchaeota genome.

According to the evolutionary model (Fig. 31.1), a major late event in eukaryogenesis was endosymbiosis of Lokiarchaeota engulfing α-proteobacteria. Lokiarchaeota was discovered deep in the mud near the Loki Trap Deep Sea Hydrothermal Vent (reported in *Nature* in 2015). Notably, Lokiarchaeota has many eukaryote-like characteristics missing in many other modern archaea. It appears that anaerobic mesophilic archaea have suffered from competition with eukaryotes and also may have been poisoned by oxygen produced by cyanobacteria. Compared to the archaea

Evolution since Coding. http://dx.doi.org/10.1016/B978-0-12-813033-9.00033-0

of ~2 billion years ago that gave rise to eukaryotes, modern archaea appear to be a story of genetic losses and streamlining to adapt to harsher niche environments. Lokiarchaeota appears to be close to one of the missing generalist archaea needed to support the model for LECA. From a scientist's point of view, the model for LECA predicted the existence of Lokiarchaeota, and now Lokiarchaeota has been found. Genes from other archaeal species appear to have been acquired in a separate nested set of earlier chimeric fusions (Fig. 31.1).

Bacterial contributions to eukaryotic genomes are similarly complex and chaotic. Some eukaryotic genes come from α-proteobacteria, but not that many. Eukaryotic introns come from bacterial group II self-splicing introns, found as single copy genes in α-proteobacteria. So, the intronic explosion in eukaryotic genomes is explained by the archaeal–α-proteobacterial fusion. The membrane constituents of mitochondria are very similar to those of α-proteobacteria. Mitochondrial DNA resembles α-proteobacterial DNA. Many other bacterial species are also represented in eukaryotic genomes, so how did their sequences get there? The nucleolus, the Golgi, and endoplasmic reticulum appear to be relics of bacterial endosymbionts. Other endosymbiotic events with other bacteria are possible, which may have contributed DNA but may not have left a recognizable cellular relic. Horizontal gene transfer occurs between archaeal and bacterial species and may also contribute DNA. Many viruses can transfer fairly large packets of genes from one organism to another. Evolution of many uniquely eukaryotic genes was driven by severe selective pressures induced during the catastrophe that was the triggering Lokiarchaeota–α-proteobacterial genome fusion. Until the presumptive eukaryote acquired mitochondria, this organism was just another archaeal–bacterial chimeric fusion. Once the mitochondria was ensconced in a complex cell, many new possibilities arose.

Chapter 34

Plants

So, where do plants come from?

As with the acquisition of mitochondria during the initial generation of eukaryotes, plants resulted from engulfment of a bacterial endosymbiont by a primitive eukaryote and subsequent genome fusion.

In this case, a photosynthetic cyanobacterium invaded a primitive eukaryote (~0.6–1.6 billion years ago).

Endosymbiotic cyanobacteria became the photosynthetic organ of algae and later of plants: the chloroplast.

Apparently, algae containing chloroplasts took up residence within other algae species. This eukaryote within eukaryote symbiosis is referred to as secondary endosymbiosis. In this way, the chloroplast/plastid was transferred horizontally to multiple, distantly (horizontally not vertically) related algae species, conferring the capacity for photosynthesis. Beautifully, this mechanism of secondary symbiosis is named *Matryoshka* (Russian nesting doll; mother doll) symbiosis. In this manner, red, purple, and brown algae were born.

At least two major endosymbiotic events are documented that generated eukaryotic subcellular organelles with fundamental functions: oxidative phosphorylation for energy generation (mitochondria; essentially all eukaryotes) and photosynthesis (chloroplasts; specific to alga and plants). Eukaryotes appear to have acquired DNA from multiple archaea and multiple bacteria from other endosymbiotic events. All eukaryotic genomes include cyanobacterial relics, even animal genomes, and animals, of course, lack photosynthesis. Cyanobacterial DNA could have been transferred to eukaryotes through other endosymbiotic events or by viral horizontal gene transfers. What endosymbiosis and horizontal gene transfer emphasize is that, to an extent, genes have a life or currency of their own and can be transferred between organisms horizontally rather than through descent. The organism, therefore, is not the only currency of evolution. With some independence, gene clusters and genes can move between organisms, particularly, perhaps, single-celled ancient organisms. Currently horizontal gene transfers occur frequently between bacteria and bacteria, between bacteria and archaea, and between archaea and archaea. Eukaryotes also engage in horizontal gene transfers and endosymbioses. The viral universe is particularly important in the horizontal transfer of genes. Plasmids and other selfish DNA agents can also function in horizontal gene transfer.

Evolution since Coding. http://dx.doi.org/10.1016/B978-0-12-813033-9.00034-2

Koonin (NCBI) tells these stories beautifully in his books and reviews.

Plants fix CO_2 using an enzyme named ribulose bisphosphate carboxylase (RuBisCo). RuBisCo is partly a TIM barrel enzyme ($\beta-\alpha)_8$, indicating its ancient ancestry. RuBisCo is not a particularly efficient enzyme for converting CO_2 into simple 3-carbon sugars (3-phosphoglycerate). Running the Calvin cycle to reduce CO_2 requires chemical energy (ATP) and redox energy (NADPH). If you expect RuBisCo to save the world from man-made global warming by reclaiming the excess CO_2 humans have excreted from the atmosphere, however, you will be disappointed. Another more aggressive method of fixing CO_2 and radical sequestration of oxidized fossil fuels will be necessary to even marginally address this existential, man-made problem.

CO_2 is a highly oxidized form of carbon (low energy on a redox (oxidation-reduction) scale). Sugars and fossil fuels are highly reduced forms of carbon (high energy on a redox scale). CO_2 is burnt carbon. Sugars are "burnt" (oxidized) through metabolism, mostly involving TIM barrel enzymes ($\beta-\alpha)_8$ and Rossmann fold enzymes ($\beta-\alpha)_8$. Burning sugars through metabolism generates chemical energy (ATP and GTP) and redox energy (NADH, NADPH). Fossil fuels are burnt (oxidized) by combination with O_2 to form CO_2. Burning fossil fuels generates energy, toxic pollution, and CO_2, a greenhouse gas.

Chapter 35

The Permian–Triassic Extinction

The biosphere can control the atmosphere.

In the lush Permian age, plants ruled the earth.

Plants living and dying accumulated a huge reservoir of dead plant matter.

Toward the end of the lush and verdant Permian age (~252 million years ago), a horizontal gene transfer event occurred between bacteria and an archaeal methanogen. Genes encoding cellulase enzymes were transferred to the methanogen to degrade plant matter (cellulose), providing a huge new store of food to a methane-generating archaea. The methanogen bloomed and CH_4 was accumulated.

Once again, the atmospheric balance of gasses was disrupted by the biosphere with devastating effects. Evolution of methane generated CO_2 via methane-eating bacteria. CH_4 and CO_2 generated heat: global warming. CO_2 dissolved in water in oceans to generate acid (H^+).

$$CO_2 + H_2O \leftrightarrow HCO_3{}^- + H^+ \leftrightarrow CO_3{}^{2-} + 2H^+$$

The advent of the Triassic Age was a time of grotesque ocean acidification causing massive extinction.

Global warming resulted in the evolution of deep-sea methane, and ~8°C of global warming may have occurred. Atmospheric CO_2 may have risen by ~2000 ppm (parts per million). Preindustrial CO_2 (survivable for humans) was ~280 ppm, for scale. Approximately 600 ppm CO_2 (which humans will soon generate) will mean death to humans on earth. Acidification of oceans by ~0.5 pH units is also death for humans, and this threshold will be passed within the next 200 years. Currently the earth (2017) is at ~400 ppm CO_2 (which is likely not survivable) and ~0.1 pH unit ocean acidification. If rampant combustion of fossil fuels is not stopped, the unrestrained burning will destroy a livable habitat for humans on earth within 50–100 years (from 2017).

Ocean acidification at the death of the Permian Age must have been unimaginable (>0.5 pH unit decrease). Because of man-made ocean acidification (far advanced in 2017), such disastrous decreases in ocean pH will soon be achieved through human activities.

There was also massive volcanic activity (the Siberian Traps) associated with this time of global extinction. If volcanic activity encountered oil or

Evolution since Coding. http://dx.doi.org/10.1016/B978-0-12-813033-9.00035-4

coal deposits, this may have enhanced the release of CO_2. There may have been asteroid impacts, but finding a 252 million year old bolide crater poses a challenge, and a characteristic iridium deposit or impact layer has not been found.

The lush Permian to desolate Triassic boundary (~252 million years ago) is referred to as the greatest known biological extinction event of all time.

Trilobites disappeared from the oceans, probably due to ocean acidification. Most crinoids disappeared.

An ~95% of all sea creatures disappeared from earth.

An ~70% of all vertebrates disappeared.

Therapsid animals mostly disappeared. After the worst of the climate and acidification bottleneck, dinosaurs eventually began to fill the void, as the initially bleak Triassic landscape began to show new signs of green.

Chapter 36

The Triassic–Jurassic Extinction

The Triassic–Jurassic Extinction event occurred ~201.3 million years ago. The Triassic–Jurassic Extinction is just prior to, and probably began, the breakup of the supercontinent named Pangaea, which ruptured into several continents known today. In keeping with the tectonic violence of the fragmenting core landmass, this was a time of immense volcanic activity, referred to as the CAMP (Central Atlantic magmatic province). Relics of the CAMP include volcanic flows hardened into 300-m-thick basalt found today, for instance, in Morocco. In addition to the rupturing of the earth, this was a time of multiple celestial bolide impacts. As I write (2017), contributions of continent rupture, volcanic violence, celestial impacts, global warming, ocean acidification, global cooling, and ocean oxygen depletion to this global disaster remains uncertain.

Many animals, sea creatures, and plants died.

Evolution since Coding. http://dx.doi.org/10.1016/B978-0-12-813033-9.00036-6

Chapter 37

The Cretaceous–Paleogene Extinction

The Permian–Triassic Extinction was the big one: ~252 million years ago. This extinction may have partly been brought on via horizontal cellulase gene transfer from bacteria to a methanogenic archaea. The biosphere controls the atmosphere.

By contrast, the Cretaceous–Paleogene Extinction was largely the result of a celestial asteroid impact upon earth ~66 million years ago. This extinction was probably not an attack of the biosphere but an attack from deeper space. The precise source and size of the asteroid is debated. The asteroid (maybe ~5–10 km in diameter) struck what is now the Gulf of Mexico near what is now Chicxulub, Yucatan, in what is now southern Mexico. The impact crater is ~180 km in diameter giving scale to the massive size of the asteroid. The impact generated a thick global deposit layer of iridium, blanketing of the earth in smoke soot and debris, unimaginable tsunamis, blocking of the sun, global cooling, destruction of forests, destruction of food for animals, and extinctions of animals. Ammonites disappeared, nonavian dinosaurs disappeared, mosasaurs and plesiosaurs disappeared, pterosaurs disappeared. An ~75% of complex organisms (eukaryotes) may have gone extinct. Other slightly smaller related asteroid impacts may have occurred at other sites, about the same time as the impact at what would become Chicxulub.

It has been estimated that the Chicxulub asteroid impact generated an explosion of ~100,000,000 megatons of TNT, essentially instantaneously.

The Deccan Traps volcanoes erupted prior to the impact and may have set the stage for subsequent megadeath upon bolide impact.

Evolution since Coding. http://dx.doi.org/10.1016/B978-0-12-813033-9.00037-8

Chapter 38

The Paleocene–Eocene Thermal Maximum

The Paleocene–Eocene Thermal Maximum occurred ~55.8 million years ago. Global temperatures rose by ~5°C. Ocean acidification caused the pH to drop catastrophically by ~0.5 pH units. The Paleocene–Eocene Thermal Maximum is an unheeded cautionary tale for current man-made global warming, which is on the cusp of recreating conditions of the Paleocene–Eocene Thermal Maximum, or worse. Currently, humans have achieved ~1°C global warming since preindustrial times. In 2017, ocean acidification is elevated by ~0.1 pH units (a 0.1 pH unit decrease; pH is a negative logarithmic scale) since pre-industrial times. In 2017, the human biosphere is existentially threatened by man-made global warming, man-made ocean acidification, and man-made sea level rise. The biosphere controls the atmosphere.

To paraphrase Pogo (Walt Kelly): *we have met the biosphere, and it is us.*

The Republican Party of the United States claims that humans are less adept at changing the environment than archaeal methanogens and cyanobacteria, much compelling evidence to the contrary.

Evolution since Coding. http://dx.doi.org/10.1016/B978-0-12-813033-9.00038-X

Chapter 39

Promoter Proximal Pausing and the CTD Interactome

The carboxy terminal domain (CTD) of RNA polymerase II evolved as a scaffold for eukaryotic evolutionary innovation (Figs. 19.5 and 32.2).

As John Stiller (East Carolina University) has indicated, the CTD likely arose to support cotranscriptional splicing of introns, which massively invaded the early eukaryotic genome as bacterial group II self-splicing introns. Genomes of the first eukaryotes were existentially threatened by group II intron explosion. The cell nucleus, CTD, and cotranscriptional splicing arose as evolutionary defenses.

The CTD is a strange repeating heptapeptide (7 amino acid repeat) sequence with the consensus sequence ^{1}YSPTSPS^7 (tyrosine1-serine2-proline3-threonine4-serine5-proline6-serine7). This repeating sequence can be heavily phosphorylated, mostly at the serine2 and serine5 positions (two SP (serine-proline) serines). Threonine4 and serine7 are also targets for phosphorylation. Phosphorylation is mostly by cyclin/cyclin-dependent kinase pairs, underscoring the similarity and coevolution of the RNA polymerase II transcription cycle and the eukaryotic cell cycle. Cyclin domains are related to helix-turn-helix (HTH) domains in general transcription factors. The CTD can be considered to be somewhat similar to a global positioning device that helps to track RNA polymerase II position in, and progression through, the transcription cycle: initiation, promoter escape, elongation, termination, and recycling. Remarkably, the eukaryotic cell cycle is similarly controlled by a set of check points that require cyclin/cyclin-dependent kinase pairs for progression. Complexity in transcription cycle control and cell cycle control, therefore, are eukaryotic innovations with a common root that diverged to support very different cyclic processes.

Reminder: archaeal/eukaryotic TFB/TFIIB and homologous bacterial σ factors (core homologous general transcription factors throughout the three domains of life) are derived in evolution from cyclin-like HTH folds.

So, the CTD is a repeat sequence generated by genetic iteration, similar to ancient α/β repeat proteins.

The CTD became a scaffold for the interaction of many factors (initially splicing factors but subsequently many other factors) that interact with RNA polymerase II to synthesize and process messenger RNA. The CTD interactome

Evolution since Coding. http://dx.doi.org/10.1016/B978-0-12-813033-9.00039-1

157

also decorates the messenger RNA for transport through the nuclear pore and licenses the messenger RNA for translation on the ribosome.

Because of the complexity of the CTD interactome, some interactome factors require consensus CTD repeats and some factors prefer or require nonconsensus (degenerate) repeats. Therefore, the CTD is not only a story of consensus repeats but also a story of degenerate repeats that coevolved with interactome factors to assume specific functions.

In complex eukaryotes, such as humans, the CTD is a combination of consensus and degenerate repeats.

In animals with complex body plans, the promoter-proximal pausing mechanism is part of the CTD interactome. For examples, *Drosophila melanogaster* (fruit fly) and humans have complex body plans and utilize promoter-proximal pausing. *Caenorhabditis elegans* (worm) has a simpler body plan and does not appear to utilize promoter-proximal pausing. Probably, with its simpler body plan and developmental process, *C. elegans* lost the promoter-proximal pausing mechanism during evolution. Plants lack promoter-proximal pausing. Therefore, a fairly complex organism (a plant or *C. elegans*) with complex development can be generated without the promoter-proximal pausing mechanism, but the most complex animals require the promoter-proximal pausing system to help direct development.

Promoter-proximal pausing allows a level of control of messenger RNA synthesis that is necessary for development of very complex animal body patterns. This feature of the CTD interactome, therefore, appears to license higher order animal complexity.

Some key players in promoter-proximal pausing include DSIF (SPT5, SPT4), NELF, cyclinT-CDK9, the PAF complex, AFF4, MLL (KMT2A), HEXIM, 7SK RNA. Very briefly, RNA polymerase II pauses transcription elongation near to the promoter and waits to be licensed to proceed through the gene. Licensing requires modification of DSIF subunit SPT5 through phosphorylation, inactivation of NELF, and activation of cyclinT-CDK9. CyclinT-CDK9 phosphorylates the CTD at serine2, which permits elongation. If promoter-proximal pausing sounds to the reader like a tangled web of add-ons and antirepression mechanisms; that is good: that is the idea.

If you search the cBioPortal cancer database (Memorial Sloan Kettering Cancer Center), you will find that many promoter-proximal pausing factors contribute to human cancers. HIV-1 (AIDS virus) transcription also subverts the promoter-proximal pausing mechanism. These observations demonstrate the expected links between cancer, viral infection, and RNA polymerase II transcription. In cancer, normal developmental pathways are subverted via accumulation of mutations (evolution) to unleash unrestrained growth of tumor cells. Viruses are evolved to subvert normal developmental pathways to support growth and/or latency of the virus. In eukaryotes, normal development and homeostasis are supported by a very complicated set of interacting networks tuned and balanced in evolution. Cancers and viral infections subvert these otherwise highly tuned, homeostatic but tangled, and cross-talking interaction systems.

The RNA polymerase II CTD and its extensive and tangled interactome are emblematic for eukaryotic complexity and evolution. These are features of gene expression control that license eukaryotic complexity and intricate genetic schemes for development of complex animals.

Chapter 40

Human Evolution

In genetic warfare and mayhem was eukarya born (apologies to Robert Ardrey). Emerging as eukaryotes did from genetic chaos, genomic and organismal complexity became the primary drivers of eukaryotic evolution.

Compared to archaea and bacteria, eukaryotes evolved into more complex cell structures, with more complex gene expression, more complex means for chemical communication, and larger genomes with more coding capacity. Because of their larger genome size, eukaryotes could maintain multiple copies of core genes, which then could diverge into related functions. For prokaryotes, by contrast, rapid replication is often so important that few genes can be maintained in multiple copies. Also, because of the mitochondria, eukaryotic cells harness and utilize much more energy than prokaryotic cells. Eukaryotes evolved complex epigenetics, histone, and chromosome modifications. Complex interacting networks coevolved to generate additional complexity. To think about eukaryotic biology, one must confront the tangled networks that are tangled for a reason. The reason is that this is the eukaryotic evolutionary survival strategy, and this is often the eukaryotic advantage: to harness energy transduction to complex regulatory and developmental networks. Therefore, one needs a network consciousness to understand eukaryotes.

Otherwise, just like every other organism on earth, humans have TIM barrels, Rossmann folds, TOPRIM domains, two double-Ψ–β-barrel-type RNA polymerases, translation elongation factor EF-G, etc. A startling family resemblance exists, therefore, between humans, other eukaryotes, archaea, and bacteria. At the level of core protein motifs, we are all part of a single family.

From the point of view of the most ancient evolution, therefore, humans are not a particularly interesting subject.

It is not difficult to imagine the evolution of the human eye or of complex human organs. Darwin mulled over the potential problem of *irreducible complexity*, but this problem is addressed through the analysis of more ancient evolution and comparative phylogenomics.

Svante Paabo has written an excellent book on human origins and radiations based on deep DNA sequencing approaches.

Apparently, there was dalliance among *Neanderthals* and *Homo* ancestors.

Denisovans, another lost tribe of *Homo*, also interbred with *Homo* ancestors and with *Neanderthals*. So *Denisovan* DNA lives in some modern humans.

Evolution since Coding. http://dx.doi.org/10.1016/B978-0-12-813033-9.00040-8

OF MICE AND MEN

So, what makes a mouse a mouse and a man a man?

The major differences between mice and men result from differences in gene expression rather than the number or the complexity of genes.

Therefore, how do differences in gene expression evolve?

One model is that enhancer elements that regulate transcription of nearby genes are moved around in evolution via transposable elements (Fig. 40.1). The human genome is littered with transposable elements, i.e., long interspersed nuclear elements (LINEs), small interspersed nuclear elements (SINEs), endogenous retroviruses (ERVs). Because these elements require retrotransposition to jump, they utilize a reverse transcriptase and integrase mechanism to copy and move from one chromosome position to occupy another. Because reverse transcriptase is harmful to cellular replication mechanisms, there is significant evolutionary pressure to reduce reverse transcriptase activity and to limit the jumping of transposable elements. SINEs lack a reverse transcriptase/integrase mechanism, so they rely on LINEs to supply this function. ERVs have reverse transcriptase/integrase. Genes encoding retrotransposition functions are tightly suppressed or may contain mutations that keep their active expression very low to restrict jumping.

About 5%–8% of the human genome is thought to be made up of ERVs. SINE-designated Alu element makes up ~11% of the human genome (~1 million copies). Because these elements can be activated to move to new positions in the human genome, albeit at a low frequency, and because these elements may carry regulatory elements for altered gene expression, human beings are

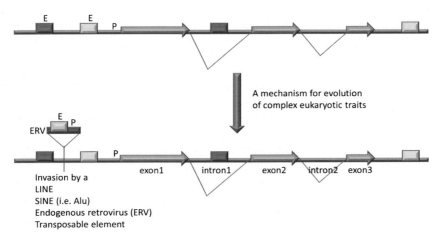

FIGURE 40.1 Human transposable elements can jump to introduce regulatory elements that alter gene expression. A human gene with 3 exons and 2 introns is indicated, with enhancer (E) and promoter (P). For instance, jumping of an endogenous retrovirus can introduce a new enhancer element, a promoter, a microRNA site, and/or another regulatory element to influence transcription and/or translation of the gene.

evolving new mechanisms for regulating genes. Much of speciation in complex eukaryotic genomes can be explained by this mechanism for altering gene regulation. One point to take from this discussion is that the human genome is dynamic, mutating, and influx. If a genome is too dynamic, the organism dies, so jumping genes are tightly regulated. MicroRNA often suppresses the activities of ERV, LINE, and SINE elements, i.e., to help reduce jumping.

Chapter 41

Human Cancer

If humans live long enough, they die of cancer, or so it is said.

Because cancers evolve through accumulation of multiple genetic changes, cancers inform about human physiology, gene regulation networks, and cellular functions. Cancers illuminate the vast coevolved networks of interrelated functions that make up a human.

Cancer cells evolve to replicate out of control, to alter normal gene expression programs, block programmed cell death, signal inappropriately within and without the cell, disrupt epigenetic programs, generate vasculature, and to move through the lymph or the blood to colonize new tumors.

Complex organismal plans, complex and interrelated biological networks, and long human life spans invite cancers.

In older humans, therefore, cancers become close to inevitable.

Cancers are generated mostly by accumulation of multiple somatic mutations.

When people speak of curing cancer, it is not clear what they are talking about. Few cancers, even of a single diagnosis, have a completely common genetic cause. Most cancers arise because of the accumulation of multiple mutations generally in a somatic (body) cell (as opposed to a germ-line cell, in which the genetic defect might be inherited by offspring). DNA sequencing of cancers shows the accumulation of many mutations, many of which may be unrelated to the presenting disease. Of course, humans can be treated for cancer and even cured. Sometimes, a cancer may return after treatment, with a new presentation, sometimes bolstered by new genetic changes. Some of those genetic changes may be induced by cancer drugs, which are often quite toxic and some are potentially mutagenic.

A quick way to learn about cancer as a genetic disease is to utilize online cancer databases. The R2 Cancer database and the cBioPortal database (Memorial Sloan Kettering Cancer Center) are useful and easy to learn. Just run the tutorials and allow browsing the site to teach you about cancer.

So, what are the causes of human cancers?

Mutations are found in (1) transcription factors; (2) signaling systems; (3) regulators of cell cycle; (4) apoptosis factors (regulators of programmed cell death); (5) suppression or inappropriate activation of developmental genes; (6) DNA repair genes; and (7) cell "immortalizing" functions. Sometimes chromosome translocations (i.e., breaking and rejoining of chromosome fragments) are

Evolution since Coding. http://dx.doi.org/10.1016/B978-0-12-813033-9.00041-X

165

partly or significantly causal in cancer. Alterations in global gene expression are typical, and epigenetic changes are hallmarks of cancers.

Because cancers are caused by accumulation of multiple mutations, ultraviolet light can induce melanomas, and radiation that causes DNA double-strand breaks can induce leukemia. DNA topoisomerase inhibitors used in cancer therapies that cause double-strand DNA breaks may cause secondary leukemias and lymphomas.

Smoking tobacco is highly toxic and carcinogenic. Carbon monoxide from tobacco smoke causes blood anoxia, inducing the transcription factor AhR/ARNT and/or the related HIF1α/ARNT, factors that are sometimes mutated or activated in cancer. Cigarette smoke generates chemicals (i.e., formaldehyde) that modify blood proteins. In 2017, ~400,000 Americans will die from cigarette smoke. These deaths represent genocide against humans by other humans motivated by greed.

Blood is ~8% of a human's body. Blood is an exceptionally important, dynamic, and delicate organ that is radically damaged by ingestion of smoke (any kind of smoke).

Metastasis (cancer spreading) is a major issue in human cancer patient survivability. This is also an issue of anoxia, because, for solid tumors to grow in new locations within the body, angiogenesis must be induced through anoxia to build new blood vessels, to bring oxygen to the growing tumor. The genes involved include AHR, ARNT, HIF1A, VEGF, and HSP90. Anticancer drugs are designed to attack the angiogenesis system to inhibit the spread of cancers. Many drugs are directed at VEGF (vascular endothelial growth factor), which is required for new blood vessels to form in response to anoxia.

So what about the RNA polymerase II carboxy terminal domain (CTD) interactome? This is a tangled web of factors linking dynamic transcriptional functions to epigenetics and to the chromatin interactome. So, what is the cancer connection? Some of the relevant genes are CDC73, CDK9, HEXIM2, GTF2A2, POLR2A, POLR2B, and NELFA. These genes may be amplified or mutated in cancers.

For mixed lineage lymphomas, there are many chromosome translocations particularly in the KMT2A gene. The relevant genes include KMT2A, CDC73, CDK9, HEXIM2, AHR, ARNT, HIF1A, and HSP90.

Most human cancers involve loss or damage to transcription factors TP53 and retinoblastoma (RB).

The RB protein is a transcriptional repressor for RNA polymerase II. In the case of the inactivation or loss of RB, it appears that loss of the repressor activates (derepresses) genes that then support unlimited growth of cells: SOX2, NANOG, KLF1, and OCT4. These are the same factors that support reprogramming of adult differentiated cells to become pluripotent stem cells. Most normal cells divide a small number of times, senesce, and die. *Immortalization* of cells allows unrestricted replication, a hallmark of cancer. Therefore, in a logical way, stem cell research interacts with cancer research, and a major contributor to human cancer (RB loss) may be explained. Loss of RB supports precancer or cancer cell immortalization.

The cBioPortal website is an easy to use, hands-on resource (that anyone can access) to learn about human evolution, cancer, and the complexity of human gene interaction networks. Gain expertise through running online tutorials. Spend a happy week feeding the website the genes listed in this chapter.

Chapter 42

Homology Modeling and Cryoelectron Microscopy

Scientists, of course, never question evolution. What is there to question? Evolution is too predictive to be ignored.

As another example of the utility of evolution, I present homology modeling. Cryoelectron microscopy is a rapidly developing method to obtain atomic resolution images of large and complex multiprotein, DNA, and/or RNA assemblies. To interpret cryoelectron microscope images, however, require construction of homology models and other approaches.

In 2017, essentially all folds of all proteins are known from existing X-ray crystal structures.

OK. So that is not quite true. Viruses are a crucible for the creation of novel protein folds. I am not certain how rapidly novel proteins generated within the rapidly evolving virus world can be introduced to the cellular world.

Regardless, limited sequence homology relating two protein sequences permits homology modeling. For instance, 20%–30% amino acid similarity among related proteins may be sufficient for modeling. Therefore, if a related protein has a known structure from cryoelectron microscopy, X-ray crystallography, or nuclear magnetic resonance, you can thread the homologous amino acid sequence to the structure, generating a highly reliable homology model.

An example of a homology model is shown in Fig. 42.1. In this case, human RNA polymerase III amino acid sequence was threaded to a homologous yeast RNA polymerase III cryoelectron microscopy image model with about 30% sequence similarity. By multiple diagnostics, the resulting human RNA polymerase III structure appears to be of similar quality to a cryoelectron microscopy image or an X-ray crystal structure. Internal hydrogen bonds and ion pairs are maintained. Everything about the structure appears stable, reasonable, reliable, and correct. Threading was done using the program Phyre2 (an online server).

It may be awhile before a cryoelectron microscopy image or an X-ray crystal structure becomes available for human RNA polymerase III. Until structural data become available, scientists can work using accurate and reliable homology models. A homology model of human RNA polymerase III is necessary, for

Evolution since Coding. http://dx.doi.org/10.1016/B978-0-12-813033-9.00042-1

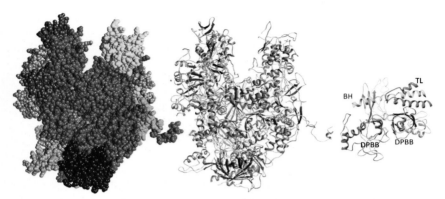

FIGURE 42.1 Homology model of human RNA polymerase III. Left image: Space-filling model, colored for different RNA polymerase III subunits. Center image: Secondary structure representation (β sheets are purple; helices are orange). Rightmost image: Detail of the active site. *BH*, bridge helix (yellow); *DPBB*, double-Ψ–β-barrel; *TL*, trigger loop (green).

instance, to understand human leukodystrophies, a recessive genetic disease that results from mutations in RNA polymerase III.

Without evolution, homology modeling would be impossible. Evolution predicts that homology modeling would be useful, and homology modeling is useful and predictive.

Using model building tools, functional and accurate structures of any multisubunit two double-Ψ–β-barrel-type RNA polymerases can be constructed from sequences based on homology and threading to known crystal structures or cryoelectron microscope images.

The theory of evolution predicts that homology modeling might be successful, and it is successful. What, if anything, does the theory of special creation predict?

Chapter 43

Human Extinction

We currently race into the Great Anthropocine Extinction (2017).

The Great Anthropocine Extinction event will compete with the Permian–Triassic Extinction of 252 million years ago: up until the present, which was the big one. The Great Anthropocine Extinction, driven by humans, will soon be the biggest catastrophe recorded in fossil and geological records.

In 2017, the human race is fully committed to disaster. We have burst past many, many critical and now irretrievable tipping points, during which the environment of earth was systematically and deliberately subverted and destroyed by humans for money. The premeditated and deliberate effort to destroy earth was supported by endless and well-funded agitation, propaganda, and lies.

Unless we can geoengineer the earth's climate thermostat and ocean acidification simultaneously, we die. This requires extracting CO_2 from the environment to maintain 200–275 ppm (parts per million) CO_2 in the atmosphere and to balance the acidity/basicity of the seas. We have no technologies to achieve such goals. Currently (2017) we are at ~400 ppm CO_2 in the atmosphere, which is neither sustainable nor survivable. Without radical geoengineering that we do not now know how to do, 400 ppm CO_2, even with no further CO_2 additions (which are inevitable), will continue to warm the planet for 1000 years because excess CO_2 is cumulative in the atmosphere. In time, smoke may dissipate and disappear to view, but CO_2 does not go away. Plants will not deplete this extra CO_2. Plants are inefficient at fixing CO_2. Plants utilize a TIM barrel α/β protein, RuBisCo (ribulose bisphosphate carboxylase; a $(\beta–\alpha)_8$ repeat protein) to fix CO_2.

Currently (2017) we are at ~1°C man-made global warming since the Industrial Revolution (mostly since the Great Acceleration (starting ~1950)). Laymen speak of 2°C warming as acceptable because 2°C warming is now the best possible case scenario anyone can imagine given the awesome power of human politics and human greed. A 4–8°C warming (death) is very likely (and probably inevitable) within the next 50–100 years. A 1°C warming (current) does not appear to be tolerable.

pH is a negative logarithmic scale. Ocean acidification is currently increased by ~0.1 pH unit (the pH of the ocean Benthic Regions has decreased by 0.1 pH unit) since the Industrial Revolution [mostly since the Great Acceleration

Evolution since Coding. http://dx.doi.org/10.1016/B978-0-12-813033-9.00043-3

(starting ~1950)]. A pH drop of 0.6–0.7 pH units is death to humanity and complex life on the planet. That is the pH drop associated with the Permian–Triassic Extinction, formerly the big one, soon to be outdone in the Great Anthropocine Extinction of today.

Man-made global warming and ocean acidification are genocide motivated by greed. Approximately, 10 million humans are dead today (2017) because of man-made global warming. By 2030, ~100 million humans will be dead due to man-made global warming. These vast underestimates come from the United Nations. Calculate morbidity to mortality at 50 to 1.

If you boil water, you drive off the dissolved CO_2 and the pH of the water increases (becomes more alkaline, approaching pH ~7). Some years ago, I did this experiment in undergraduate chemistry laboratory at UCLA. Today (2017), I see no reason to question that CO_2 dissolves in water to produce acid. Even as an undergraduate, I believed this, and my chemistry professor was also convinced and indeed was quite emphatic. It appears that some in America deny this fact for very short-term political and monetary gain.

It is interesting to consider whether ongoing, radical, man-made purifying selection is "natural" selection or just dumb imbecilic mass suicide: the stuff of which America is made.

Some may judge this chapter as polemics, but, I assure you, it is not. The subject of this book is evolution. Man-made global warming and man-made ocean acidification are purifying selection to remedy excesses of a very recent and ultimately short-lived species.

The historic Paris Accord (COP21) strongly opposed by Donald J Trump is nowhere near sufficient to significantly alter this tragic trajectory.

Chapter 44

Evolution Versus Faith

Antievolution propaganda is everywhere in the United States of America, but, why this should be so or be an issue, is unclear. Many Americans favor teaching radical propaganda in preference to teaching biology and evolution.

Evolution is in no way an antireligious concept.

Evolution has nothing to do with religion.

Using science to falsify the existence of a deity would be odd and pointless.

Using holy writs to falsify science seems similarly stupid and pointless. Do the devout really want the Bible or Koran to be a science text? As such, these works would be somewhat out of date.

Evolution is no more antireligious or anti-Bible than jet airplanes and wristwatches.

Evolution describes biology as does nothing else.

Intelligent design, by contrast, does not describe biology, speciation, gene regulation, ancient evolution, or cancer.

Evolution provides a detailed, comprehensive, and largely testable working model for the most ancient genesis of life on earth. Very few details appear to be missing from this amazing but coherent story. Every major event is described in detail and backed up with supporting evidence and data that anyone with curiosity can access. In the interest of brevity and storytelling, only some of these data and evidence are presented in this book. Going forward, it is expected that the story will continue to grow and improve in concept and detail.

Remarkably, relics of the most ancient life forms are preserved on earth. Transfer RNA and ribosome are relics of the RNA-protein world. Two double-Ψ–β-barrel RNA polymerases, RIFT barrel proteins, translation elongation factor EF-Tu, TIM barrels, Rossmann folds, kinases, and ABC ATPases are relics of the ancient RNA-protein world. By any human timescale, these protein folds are immortal, almost 4 billion years old. Folds were generated as a genetic cutout doll problem of motif repetition and folding. These ancient protein folds existed on earth before cells, which date to LUCA.

Relics of LUCA include all of the above, helix-turn-helix motifs, winged helix-turn-helix motifs, DNA-template-dependent two double-Ψ–β-barrel RNA polymerases, TATA-binding protein, σ factors, TFB, TATA boxes, BREs, and Pribnow boxes. Many DNA repair proteins trace to LUCA.

Evolution since Coding. http://dx.doi.org/10.1016/B978-0-12-813033-9.00044-5

Divergence of archaea and bacteria from LUCA probably generated modern DNA replication systems separately in the two lineages.

The first eukaryote was an endosymbiotic fusion of an ancient Lokiarchaeota archaea and α-proteobacteria. This violent collision of genomes resulted in a more complex organism with organelles, genes broken by introns, and many new and self-induced selective pressures. As a defense against translation of intron RNA the nucleus became necessary. After fusion, the archaeal anaerobe was capable of utilizing oxygen to drive metabolism. Oxygen accumulated in the environment because it was excreted by photosynthetic cyanobacteria. It was not clear previously whether the anaerobic, generalist, mesophilic archaea that gave rise to eukaryotes is now extinct or just difficult to find. Deeply buried archaea of the Lokiarchaeota family, so far, appear most similar to eukaryotes, and, therefore, the most likely endosymbiotic fusion partners with an α-proteobacterium. The advent of eukaryotic organisms with mitochondria, a nucleus, introns, and novel genetics and cellular complexity generated fierce competition for their parental archaea.

As survival strategies, eukaryotes developed network and genomic complexity. Integrated and tangled network systems coevolved for gene expression, chromatin, cell cycle, signaling, and epigenetics. These more complex network systems and more nuanced gene regulation capabilities encouraged multicellularity.

The carboxy terminal domain (CTD) of eukaryotic RNA polymerase II evolved initially to support splicing of introns, introduced from the invading α-proteobacteria into the archaeal genome. The CTD became a scaffold for recruitment and evolution of a vast network of factors to control cotranscriptional processes. The transcription cycle of initiation, promoter escape, elongation, termination, and recycling resembles a cell cycle, and the transcription cycle and the cell cycle must be initially coevolved in eukaryotes. The vast CTD interactome interfaces to the chromatin interactome. Disruption of these powerfully interacting systems is a hallmark of cancer. Intricate networks are typical of eukaryotic biology, which, because of eukaryote complexity, invites disruptions such as cancers and viral infections. Specific evolution of factors within the CTD interactome has licensed increased complexity in developmental gene expression programs that correlate with increased complexity in the body plans of animals. Analogous to α/β proteins, the CTD is an example of motif repetition. In the case of the CTD, a complex network was recruited to, and evolved around, the new scaffold.

Nothing could be more obvious or visible than evolution. Evolution is no more antireligious or anti-Bible than jet airplanes and wristwatches.

What the study of evolution and protein structure seems to do is to make a deity irrelevant, at least for the genesis of life on planet earth.

Chapter 45

Concept in Biology

So, does biology have concept or is it *stamp collecting*?

By stamp collecting, I refer to collection of seemingly infinite minutiae and factoids, such as those accumulated in genomics research and deposited in seemingly endless databases. I do not mean to criticize genome research, which I honor and struggle to utilize, but the output can be daunting.

This book was intended as a means to develop and eventually to teach increased concept in biology. In this book, I am striving toward working models that make biology, which is perhaps too often taught as stamp collecting, easier to understand.

In education, always motivate the student first and keep the motivation positive rather than punitive. Teachers organize and present material and aim to inspire students. Students, sadly, must teach themselves the material, so ultimately the student is the best teacher, and the teacher has another role. Students learn what teachers inspire them to learn, or students with more developed intellectual skills simply teach themselves material that they become motivated to learn on their own. Teachers can be necessary and useful as learning coaches, but, often, their role does not appear to be properly credited or understood. Concepts, oddly, can be more motivating to students than infinite factoids.

Computers are a powerful tool to learn biochemistry, genetics, protein structure, and evolution. Learning molecular graphics is a powerful approach to integrate protein structure, evolution, and dynamics. Cancer databases are powerful learning tools. When computers are properly used and coached, they become tools for effective self-teaching.

Concept in biology only comes in retrospect through evolution and purifying selection. There does not appear to be an intelligent design going forward, and resulting biological networks and designs do not appear intelligent. Rather, existing biological designs appear survivable.

Truly, evolution is the essential, core concept in biology.

This book presents a simple and straightforward working model and story for genesis of life on earth, told c.2017. The story will continue to evolve and improve, but significant detail and concept are already embedded in the tale. I advocate using these concepts to more effectively integrate the teaching of biology, evolution, and protein structure/function/dynamics.

Evolution since Coding. http://dx.doi.org/10.1016/B978-0-12-813033-9.00045-7

To me, the entire story makes frightening, amazing, and startling sense. A few years ago, c.2013, I could not have imagined this book, these models, or concepts. At that time, I was naive about ancient evolution and its link to RNA polymerases and general transcription factors. Now, I consider the ancient evolution field to have moved to a comprehensive understanding and integration.

Evolution provides concept to the study of biology.

Chapter 46

Other Books and Studies

To improve the storytelling in *Evolution Since Coding* references were not included with the text. Because, up to this chapter, the book lacks references, I offer this chapter as a guide to other readings.

Ideas and models expressed here are not necessarily my own. Many others have contributed to this story, and, in some ways, I remain a minor contributor. I began this book because my work on multisubunit RNA polymerases of the two double-Ψ–β-barrel type, and, also, my work on general transcription factors and promoters seemed to demand it (Burton, 2014; Burton and Burton, 2014; Wang et al., 2013, 2015; Nedialkov and Burton, 2013; Nedialkov et al., 2012, 2013; Kireeva et al., 2012; Feig and Burton, 2010; Seibold et al., 2010). When I was working on *Evolution Since Coding*, much to my surprise, I figured out transfer RNA evolution (Pak et al., 2017; Root-Bernstein et al., 2016). Of course, at first, I thought tRNA evolution and translation were intractable problems, and, to my sorrow (2017), I still do not know clearly how ribosomal RNA evolved or how the genetic code sectored into a code for 20 amino acids and stop codons. Making the unexpected discovery about tRNA evolution, however, demanded adding translation to a book that was initially centered on transcription (Pak et al., 2017; Root-Bernstein et al., 2016; Burton et al., 2016; Burton and Burton, 2014). I knew when I started, of course, that translation was the elephant in the room that I was overlooking. Because of the wearying complexity of biological systems, scientists jealously maintain blind spots. If you understand transcription and translation, however, you understand ancient evolution from the RNA-protein world forward, so a book that considers both processes is much more satisfactory. Also, when I was younger and struggling to become something resembling a scientist, I was deeply influenced by science books, and I wanted my book to be a potential inspiration.

Some of my favorite popular books on science and molecular biology include: *The Double Helix* (James Watson), *The Eighth Day of Creation* (Horace Judson) and *A Genetic Switch* (Mark Ptashne). I still have an ancient paperback of *The Double Helix* that I read as a teenager, but did not understand very well because I had not yet taken much organic chemistry. I would like to say Watson's book inspired me at my first reading, but, mostly, it confused me. I consider Watson and Crick's discovery of the DNA double helix to be one of the great intellectual leaps to occur during my lifetime. Their story remains a lesson in how to engage and solve an important biological problem. On the

Evolution since Coding. http://dx.doi.org/10.1016/B978-0-12-813033-9.00046-9

other hand, I would caution that Watson's book ought not to be taken as a guide for appropriate interactions with female colleagues. In my experience, women are better treated today by colleagues than was the case for Rosalind Franklin, although scientists of all genders face stiff suppression from society.

Molecular graphics images have been made with Pymol (www.pymol. org), YASARA (www.yasara.org), Visual Molecular Dynamics (VMD; http://www.ks.uiuc.edu/Research/vmd/), and UCSF Chimera (https://www. cgl.ucsf.edu/chimera/download.html). Learning these essential tools allows you to teach yourself molecular biology, chemistry, evolution, and art.

Protein Data Bank (PDB) codes are a means to find references for studies. The relevant references describing structures can be obtained through the RCSB Protein Data Bank.

On evolution, *Logic of Chance* (Eugene Koonin) (http://www.ncbi.nlm. nih.gov/CBBresearch/Koonin/) is great and goes into wonderful and rich details on the Virus World and different models for LUCA and LECA. Koonin's book is written from a much different point of view than my book, and also Koonin's lovely book is written from his extensive evolutionary and phylogenomic expertise, which I lack. Koonin has amazing insights into viruses and viral evolution. If you wish to read a beautiful and inspiring scientific study, read the Koonin review (Koonin, 2006, 2014, 2015a,b). My book is more of a *how to* book for the cutout doll evolution of life. I try to suggest that if you were a deity using evolutionary genetic tools, this is how you would do it. I remain overwhelmed and awed by the concept that, despite immense antiquity, the RNA-protein world, LUCA, and LECA can be visualized with such clarity. Some of my colleagues, amazingly, remain stoic in the knowledge and presence of ~4 billion year old protein folds. Just to be clear: this is *billion* not *million*, and *4 billion years* is both a long time and not a misprint.

Nick Lane has a recent (2015) book on ancient evolution *The Vital Question*. His book deals with cell energetics in great detail. He also gives interesting working models for evolution of LUCA and divergence of archaea and bacteria. He gives interesting insights into evolution of eukaryotic complexity. Compared to mine, his book is not as focused on protein structural motifs, metabolic enzymes, transcriptional mechanisms, and translation mechanisms. My book is more about breaking the ancient codes written in protein and RNA sequences and structures (Pak et al., 2017; Root-Bernstein et al., 2016; Burton et al., 2016; Burton, 2014; Burton and Burton, 2014).

Jack Szostak (Harvard University) has written many books about protocells (Blain and Szostak, 2014). I look forward to reading more of these. In my book, I do not treat the generation of cells in great detail because I concentrate on RNA synthesis mechanisms, tRNA evolution, and their links to coding (Pak et al., 2017; Root-Bernstein et al., 2016; Burton et al., 2016; Burton, 2014; Burton and Burton, 2014). In order to generate energy by maintaining a proton or ion gradient across a membrane to drive redox reactions, cells are necessary. Closed membrane systems, therefore, remarkably, must predate intact cellular organisms with streamlined DNA genomes, which arose at LUCA.

Andrei Lupas and colleagues (Max Planck Institute) (Alva et al., 2008, Coles et al., 1999, 2005, 2006; Soding and Lupas, 2003) determined pathways for evolution of cradle-loop barrel folds: RIFT barrels, double-Ψ–β-barrels, and swapped hairpin barrels.

The story of LUCA evolving in a soup of bubbling hydrogen and carbon dioxide gas is remarkable and inspiring (Weiss et al., 2016).

Koonin (Rogozin et al., 2012; Koonin, 2006, 2014) and Baum and Baum (2014) describe the birth of the cell nucleus. These are wonderful stories. These stories, however, may be replaced by a complex story of *Matryoshka* (Russian nesting doll) endosymbiosis (Pittis and Gabaldon, 2016), as described in Chapter 31. Currently, this is the story that I favor for LECA and FECA→LECA.

The remarkable story of Lokiarchaeota in LECA was recently told (Zaremba-Niedzwiedzka et al., 2017; Spang et al., 2015; Pittis and Gabaldon, 2016).

Koonin and colleagues explained the birth of introns, ubiquitous and widely conserved in eukaryotic genomes (Rogozin et al., 2012; Koonin, 2009). Some simple eukaryotes lacking many introns appear to have lost them in evolution. The story of intron evolution in eukaryotes is wondrous and exciting.

Aravind and Koonin explained evolution of two double-Ψ–β-barrel type RNA polymerases (Iyer and Aravind, 2012; Iyer et al., 2003). My interest in these enzymes helped me see the importance of taking the evolutionary view, because multisubunit RNA polymerases tell such a rich story of genesis. When I applied simple evolutionary concepts to multisubunit RNA polymerases, their exquisite dynamics and complex structures suddenly made sense (Burton et al., 2016; Burton, 2014; Wang et al., 2013). Previously, I was ignorant and confused.

John Stiller reported on the evolution of the RNA polymerase II CTD, an essential feature specific to and defining for eukaryotes. This is a wonderful study (Yang and Stiller, 2014).

John Stiller also described *Matryoshka* secondary endosymbiosis in transfer of photosynthetic plastids (Bodyl et al., 2009; Stiller et al., 2009; Stiller, 2007). I had not appreciated eukaryote to eukaryote endosymbioses previously. Remarkably, *Matryoshka* endosymbiosis appears to describe eukaryogenesis, as well (Pittis and Gabaldon, 2016). Some ancient archaea were capable of engulfing an endosymbiont (Zaremba-Niedzwiedzka et al., 2017; Spang et al., 2015).

Dirk Eick, Matthias Geyer, Francois Robert, and Jeffrey Corden helped me to partly understand the RNA polymerase II CTD (Eick and Geyer, 2013; Jeronimo et al., 2013; Corden, 2013).

Robert Weinzierl (2013) and Finn Werner (Blombach et al., 2013, 2015; Nagy et al., 2015; Werner, 2013) helped me to understand archaeal RNA polymerase and archaeal general transcription factors.

Judith Jaehning (Shi et al., 1997; Wade et al., 1996) and Karen Arndt (Tomson and Arndt, 2013; Crisucci and Arndt, 2011; Arndt and Kane, 2003) helped me to partly understand and appreciate the PAF complex. With Judith Jaehning, I did some early work to identify the PAF complex in yeast (Shi et al., 1997; Wade et al., 1996).

Karen Adelman (NIH) (Williams et al., 2015; Rogatsky and Adelman, 2014; Henriques et al., 2013; Fromm et al., 2013) and John Lis (Cornell University) (Jonkers and Lis, 2015; Buckley et al., 2014; Burgess, 2013; Fuda and Lis, 2013; Saunders et al., 2013) helped me to understand promoter-proximal pausing by RNA polymerase II in complex animals.

Seth Darst (Rockefeller University) (Lane and Darst, 2010a,b) started me thinking about RNA polymerase evolution. I saw Darst give a talk on this subject at a meeting in Chicago (Summer, 2013). Up until his talk, I had not thought much about ancient evolution or its importance. I had no idea that this subject was so far advanced or was so relevant to understanding multisubunit RNA polymerases (Burton et al., 2016; Burton, 2014), their dynamics and their general transcription factors (Burton et al., 2016; Burton, 2014; Burton and Burton, 2014). I think I understand better now. I hope I have related this amazing and intricate story in an accessible and interesting way.

Jack Greenblatt (University of Toronto) (Ni et al., 2011, 2014; Zhou et al., 2009; Greenblatt, 2008) helped me to understand the tangled mess that is eukaryotic transcription. I have long considered eukaryotic transcription as a kind of implausible Rube Goldberg device lacking a sense of humor. Cobbled together as it is with evolutionary bailing wire and chewing gum, eukaryotic transcription appears to be a compelling argument against intelligent design.

Dick Burgess (University of Wisconsin) (Burton et al., 1981, 1983; Burgess et al., 1969) and Carol Gross (University of California, San Francisco) (Feklistov et al., 2014) introduced me to bacterial σ-factors.

Jim Ingles (University of Toronto) (Moyle et al., 1989; Allison et al., 1988) introduced me to the RNA polymerase II CTD.

Svante Paabo has written a book about human evolution (*Neanderthal Man: In Search for Lost Genomes*). He knows much more about this subject than I do. In my book, I give the subject of human evolution somewhat short shrift. I do not find human evolution particularly unique, so I am the wrong person to expound on this subject, and it was already done. I am trying to look at older events and times.

I do touch on human extinction, which I find interesting but immensely frightening. One would rather leave a habitable world for one's children and grandchildren. It appears now that a habitable earth for our children is not politically feasible (2017).

In a book about evolution, Charles Darwin must be mentioned. Although Darwin knew nothing about molecular mechanisms, his theories adapt surprisingly well to a molecular view. I do not understand the intense and bitter hatred directed against Darwin by antievolutionists. For instance, our local throwaway newspaper published a commentary that described Darwin as a "God hater." I doubt Darwin hated god. Personally, I do not think evolution has anything to do with God, certainly, no more than jet engines and wristwatches. I admire Darwin's enduring relevance to an intense modern debate and discussion.

I am not certain that I got all the details exactly right for various ancient asteroid impacts and extinctions. In 2017, new ideas and details bombard one's consciousness, and this is not my focus of study, so I just soak it in as best I can as an interested amateur. I hope, however, to convey the violence and chaos that was the backdrop for the evolution of life (i.e., the Early and Late Heavy Bombardments, Snowball Earth, and the Cretaceous-Paleogene Extinction). I am certain I captured the 2017 extinction of humans accurately, and this part of the story will come to pass essentially as described. Only the precise timeline for human extinction is not completely certain, because the inexorable process is fully in place. The historic Paris Accord (COP21; 2015) will not be sufficient to alter the fate of humanity. COP21 was a baby step in a more positive direction, soon to be undone through bitter and criminal American politics.

REFERENCES

Allison, L.A., Wong, J.K., Fitzpatrick, V.D., Moyle, M., Ingles, C.J., 1988. The C-terminal domain of the largest subunit of RNA polymerase II of *Saccharomyces cerevisiae, Drosophila melanogaster*, and mammals: a conserved structure with an essential function. Mol. Cell Biol. 8, 321–329.

Alva, V., Koretke, K.K., Coles, M., Lupas, A.N., 2008. Cradle-loop barrels and the concept of metafolds in protein classification by natural descent. Curr. Opin. Struct. Biol. 18, 358–365.

Arndt, K.M., Kane, C.M., 2003. Running with RNA polymerase: eukaryotic transcript elongation. Trends Genet. 19, 543–550.

Baum, D.A., Baum, B., 2014. An inside-out origin for the eukaryotic cell. BMC Biol. 12, 76.

Blain, J.C., Szostak, J.W., 2014. Progress toward synthetic cells. Annu. Rev. Biochem. 83, 615–640.

Blombach, F., Daviter, T., Fielden, D., Grohmann, D., Smollett, K., Werner, F., 2013. Archaeology of RNA polymerase: factor swapping during the transcription cycle. Biochem. Soc. Trans. 41, 362–367.

Blombach, F., Salvadori, E., Fouqueau, T., Yan, J., Reimann, J., Sheppard, C., Smollett, K.L., Albers, S.V., Kay, C.W., Thalassinos, K., Werner, F., 2015. Archaeal TFEalpha/beta is a hybrid of TFIIE and the RNA polymerase III subcomplex hRPC62/39. Elife 4.

Bodyl, A., Mackiewicz, P., Stiller, J.W., 2009. Early steps in plastid evolution: current ideas and controversies. Bioessays 31, 1219–1232.

Buckley, M.S., Kwak, H., Zipfel, W.R., Lis, J.T., 2014. Kinetics of promoter Pol II on Hsp70 reveal stable pausing and key insights into its regulation. Genes Dev. 28, 14–19.

Burgess, D.J., 2013. Gene expression: time flies thanks to Pol II pausing. Nat. Rev. Genet. 14, 441.

Burgess, R.R., Travers, A.A., Dunn, J.J., Bautz, E.K., 1969. Factor stimulating transcription by RNA polymerase. Nature 221, 43–46.

Burton, S.P., Burton, Z.F., 2014. The sigma enigma: bacterial sigma factors, archaeal TFB and eukaryotic TFIIB are homologs. Transcription 5, e967599.

Burton, Z., Burgess, R.R., Lin, J., Moore, D., Holder, S., Gross, C.A., 1981. The nucleotide sequence of the cloned rpoD gene for the RNA polymerase sigma subunit from *E. coli* K12. Nucleic Acids Res. 9, 2889–2903.

Burton, Z.F., 2014. The Old and New Testaments of gene regulation. Evolution of multi-subunit RNA polymerases and co-evolution of eukaryote complexity with the RNAP II CTD. Transcription 5, e28674.

Burton, Z.F., Gross, C.A., Watanabe, K.K., Burgess, R.R., 1983. The operon that encodes the sigma subunit of RNA polymerase also encodes ribosomal protein S21 and DNA primase in *E. coli* K12. Cell 32, 335–349.

Burton, Z.F., Opron, K., Wei, G., Geiger, J.H., 2016. A model for genesis of transcription systems. Transcription 7, 1–13.

Coles, M., Diercks, T., Liermann, J., Groger, A., Rockel, B., Baumeister, W., Koretke, K.K., Lupas, A., Peters, J., Kessler, H., 1999. The solution structure of VAT-N reveals a 'missing link' in the evolution of complex enzymes from a simple betaalphabetabeta element. Curr. Biol. 9, 1158–1168.

Coles, M., Djuranovic, S., Soding, J., Frickey, T., Koretke, K., Truffault, V., Martin, J., Lupas, A.N., 2005. AbrB-like transcription factors assume a swapped hairpin fold that is evolutionarily related to double-psi beta barrels. Structure 13, 919–928.

Coles, M., Hulko, M., Djuranovic, S., Truffault, V., Koretke, K., Martin, J., Lupas, A.N., 2006. Common evolutionary origin of swapped-hairpin and double-psi beta barrels. Structure 14, 1489–1498.

Corden, J.L., 2013. RNA polymerase II C-terminal domain: tethering transcription to transcript and template. Chem. Rev. 113, 8423–8455.

Crisucci, E.M., Arndt, K.M., 2011. The roles of the Paf1 complex and associated histone modifications in regulating gene expression. Genet. Res. Int. 2011.

Eick, D., Geyer, M., 2013. The RNA polymerase II carboxy-terminal domain (CTD) code. Chem. Rev. 113, 8456–8490.

Feig, M., Burton, Z.F., 2010. RNA polymerase II with open and closed trigger loops: active site dynamics and nucleic acid translocation. Biophys. J. 99, 2577–2586.

Feklistov, A., Sharon, B.D., Darst, S.A., Gross, C.A., 2014. Bacterial sigma factors: a historical, structural, and genomic perspective. Annu. Rev. Microbiol. 68.

Fromm, G., Gilchrist, D.A., Adelman, K., 2013. SnapShot: transcription regulation: pausing. Cell 153 930–930e1.

Fuda, N.J., Lis, J.T., 2013. A new player in Pol II pausing. EMBO J. 32, 1796–1798.

Greenblatt, J.F., 2008. Transcription termination: pulling out all the stops. Cell 132, 917–918.

Henriques, T., Gilchrist, D.A., Nechaev, S., Bern, M., Muse, G.W., Burkholder, A., Fargo, D.C., Adelman, K., 2013. Stable pausing by RNA polymerase II provides an opportunity to target and integrate regulatory signals. Mol. Cell 52, 517–528.

Iyer, L.M., Aravind, L., 2012. Insights from the architecture of the bacterial transcription apparatus. J. Struct. Biol. 179, 299–319.

Iyer, L.M., Koonin, E.V., Aravind, L., 2003. Evolutionary connection between the catalytic subunits of DNA-dependent RNA polymerases and eukaryotic RNA-dependent RNA polymerases and the origin of RNA polymerases. BMC Struct. Biol. 3, 1.

Jeronimo, C., Bataille, A.R., Robert, F., 2013. The writers, readers, and functions of the RNA polymerase II C-terminal domain code. Chem. Rev. 113, 8491–8522.

Jonkers, I., Lis, J.T., 2015. Getting up to speed with transcription elongation by RNA polymerase II. Nat. Rev. Mol. Cell Biol. 16, 167–177.

Kireeva, M.L., Opron, K., Seibold, S.A., Domecq, C., Cukier, R.I., Coulombe, B., Kashlev, M., Burton, Z.F., 2012. Molecular dynamics and mutational analysis of the catalytic and translocation cycle of RNA polymerase. BMC Biophys. 5, 11.

Koonin, E.V., 2006. The origin of introns and their role in eukaryogenesis: a compromise solution to the introns-early versus introns-late debate? Biol. Direct 1, 22.

Koonin, E.V., 2009. Intron-dominated genomes of early ancestors of eukaryotes. J. Hered. 100, 618–623.

Koonin, E.V., 2014. The origins of cellular life. Antonie Van Leeuwenhoek 106, 27–41.

Koonin, E.V., 2015a. Archaeal ancestors of eukaryotes: not so elusive any more. BMC Biol. 13, 84.

Koonin, E.V., 2015b. Origin of eukaryotes from within archaea, archaeal eukaryome and bursts of gene gain: eukaryogenesis just made easier? Philos. Trans. R. Soc. Lond. B Biol. Sci. 370.

Lane, W.J., Darst, S.A., 2010a. Molecular evolution of multisubunit RNA polymerases: sequence analysis. J. Mol. Biol. 395, 671–685.

Lane, W.J., Darst, S.A., 2010b. Molecular evolution of multisubunit RNA polymerases: structural analysis. J. Mol. Biol. 395, 686–704.

Moyle, M., Lee, J.S., Anderson, W.F., Ingles, C.J., 1989. The C-terminal domain of the largest subunit of RNA polymerase II and transcription initiation. Mol. Cell Biol. 9, 5750–5753.

Nagy, J., Grohmann, D., Cheung, A.C., Schulz, S., Smollett, K., Werner, F., Michaelis, J., 2015. Complete architecture of the archaeal RNA polymerase open complex from single-molecule FRET and NPS. Nat. Commun. 6, 6161.

Nedialkov, Y.A., Burton, Z.F., 2013. Translocation and fidelity of *Escherichia coli* RNA polymerase. Transcription 4, 136–143.

Nedialkov, Y.A., Nudler, E., Burton, Z.F., 2012. RNA polymerase stalls in a post-translocated register and can hyper-translocate. Transcription 3, 260–269.

Nedialkov, Y.A., Opron, K., Assaf, F., Artsimovitch, I., Kireeva, M.L., Kashlev, M., Cukier, R.I., Nudler, E., Burton, Z.F., 2013. The RNA polymerase bridge helix YFI motif in catalysis, fidelity and translocation. Biochim. Biophys. Acta 1829, 187–198.

Ni, Z., Olsen, J.B., Guo, X., Zhong, G., Ruan, E.D., Marcon, E., Young, P., Guo, H., Li, J., Moffat, J., Emili, A., Greenblatt, J.F., 2011. Control of the RNA polymerase II phosphorylation state in promoter regions by CTD interaction domain-containing proteins RPRD1A and RPRD1B. Transcription 2, 237–242.

Ni, Z., Xu, C., Guo, X., Hunter, G.O., Kuznetsova, O.V., Tempel, W., Marcon, E., Zhong, G., Guo, H., Kuo, W.H., Li, J., Young, P., Olsen, J.B., Wan, C., Loppnau, P., Bakkouri M, E.L., Senisterra, G.A., He, H., Huang, H., Sidhu, S.S., Emili, A., Murphy, S., Mosley, A.L., Arrowsmith, C.H., Min, J., Greenblatt, J.F., 2014. RPRD1A and RPRD1B are human RNA polymerase II C-terminal domain scaffolds for Ser5 dephosphorylation. Nat. Struct. Mol. Biol. 21, 686–695.

Pak, D., ROOT-Bernstein, R., Burton, Z.F., 2017. tRNA structure and evolution and standardization to the three nucleotide genetic code. Transcription 20.

Pittis, A.A., Gabaldon, T., 2016. Late acquisition of mitochondria by a host with chimaeric prokaryotic ancestry. Nature 531, 101–104.

Rogatsky, I., Adelman, K., 2014. Preparing the first responders: building the inflammatory transcriptome from the ground up. Mol. Cell 54, 245–254.

Rogozin, I.B., Carmel, L., Csuros, M., Koonin, E.V., 2012. Origin and evolution of spliceosomal introns. Biol. Direct 7, 11.

Root-Bernstein, R., Kim, Y., Sanjay, A., Burton, Z.F., 2016. tRNA evolution from the proto-tRNA minihelix world. Transcription 7, 153–163.

Saunders, A., Core, L.J., Sutcliffe, C., Lis, J.T., Ashe, H.L., 2013. Extensive polymerase pausing during Drosophila axis patterning enables high-level and pliable transcription. Genes Dev. 27, 1146–1158.

Seibold, S.A., Singh, B.N., Zhang, C., Kireeva, M., Domecq, C., Bouchard, A., Nazione, A.M., Feig, M., Cukier, R.I., Coulombe, B., Kashlev, M., Hampsey, M., Burton, Z.F., 2010. Conformational coupling, bridge helix dynamics and active site dehydration in catalysis by RNA polymerase. Biochim. Biophys. Acta 1799, 575–587.

Shi, X., Chang, M., Wolf, A.J., Chang, C.H., Frazer-Abel, A.A., Wade, P.A., Burton, Z.F., Jaehning, J.A., 1997. Cdc73p and Paf1p are found in a novel RNA polymerase II-containing complex distinct from the Srbp-containing holoenzyme. Mol. Cell Biol 17, 1160–1169.

Soding, J., Lupas, A.N., 2003. More than the sum of their parts: on the evolution of proteins from peptides. Bioessays 25, 837–846.

Spang, A., Saw, J.H., Jorgensen, S.L., Zaremba-Niedzwiedzka, K., Martijn, J., Lind, A.E., Van Eijk, R., Schleper, C., Guy, L., Ettema, T.J., 2015. Complex archaea that bridge the gap between prokaryotes and eukaryotes. Nature 521, 173–179.

Stiller, J.W., 2007. Plastid endosymbiosis, genome evolution and the origin of green plants. Trends Plant Sci. 12, 391–396.

Stiller, J.W., Huang, J., Ding, Q., Tian, J., Goodwillie, C., 2009. Are algal genes in nonphotosynthetic protists evidence of historical plastid endosymbioses? BMC Genomics 10, 484.

Tomson, B.N., Arndt, K.M., 2013. The many roles of the conserved eukaryotic Paf1 complex in regulating transcription, histone modifications, and disease states. Biochim. Biophys. Acta 1829, 116–126.

Wade, P.A., Werel, W., Fentzke, R.C., Thompson, N.E., Leykam, J.F., Burgess, R.R., Jaehning, J.A., Burton, Z.F., 1996. A novel collection of accessory factors associated with yeast RNA polymerase II. Protein Expr. Purif. 8, 85–90.

Wang, B., Opron, K., Burton, Z.F., Cukier, R.I., Feig, M., 2015. Five checkpoints maintaining the fidelity of transcription by RNA polymerases in structural and energetic details. Nucleic Acids Res. 43, 1133–1146.

Wang, B., Predeus, A.V., Burton, Z.F., Feig, M., 2013. Energetic and structural details of the trigger-loop closing transition in RNA polymerase II. Biophys. J. 105, 767–775.

Weinzierl, R.O., 2013. The RNA polymerase factory and archaeal transcription. Chem. Rev. 113, 8350–8376.

Weiss, M.C., Sousa, F.L., Mrnjavac, N., Neukirchen, S., Roettger, M., NELSON-Sathi, S., Martin, W.F., 2016. The physiology and habitat of the last universal common ancestor. Nat. Microbiol. 1, 16116.

Werner, F., 2013. Molecular mechanisms of transcription elongation in archaea. Chem. Rev. 113, 8331–8349.

Williams, L.H., Fromm, G., Gokey, N.G., Henriques, T., Muse, G.W., Burkholder, A., Fargo, D.C., Hu, G., Adelman, K., 2015. Pausing of RNA polymerase II regulates mammalian developmental potential through control of signaling networks. Mol. Cell 58, 311–322.

Yang, C., Stiller, J.W., 2014. Evolutionary diversity and taxon-specific modifications of the RNA polymerase II C-terminal domain. Proc. Natl. Acad. Sci. U.S.A. 111, 5920–5925.

Zaremba-Niedzwiedzka, K., Caceres, E.F., Saw, J.H., Backstrom, D., Juzokaite, L., Vancaester, E., Seitz, K.W., Anantharaman, K., Starnawski, P., Kjeldsen, K.U., Stott, M.B., Nunoura, T., Banfield, J.F., Schramm, A., Baker, B.J., Spang, A., Ettema, T.J., 2017. Asgard archaea illuminate the origin of eukaryotic cellular complexity. Nature 541, 353–358.

Zhou, K., Kuo, W.H., Fillingham, J., Greenblatt, J.F., 2009. Control of transcriptional elongation and cotranscriptional histone modification by the yeast BUR kinase substrate Spt5. Proc. Natl. Acad. Sci. U.S.A. 106, 6956–6961.

Supplementary Materials

1. **Supplementary Material-01**
   Human TFIIB Archaeal TFB and Bacterial Sigma Factors Are Homologs (see PowerPoint file)
   tRNA as a molecular archetype (see PowerPoint file)

2. **Supplementary Material-02**
   β–α repeats: metabolism (see PowerPoint file)

3. **Supplementary Material-03**
   Double-ψ–β-barrels (see PowerPoint file)
   Evolution of the three domains (see PowerPoint file)
   Human TFIIB Archaeal TFB and Bacterial Sigma Factors Are Homologs (see PowerPoint file)

4. **Supplementary Material-04**
   Hemoglobin (see PowerPoint file)

5. **Supplementary Material-05**
   Hemoglobin Movie (see video file)

Note : PowerPoint / Video related to this materials can be accessed in the online version of the book.

Index

'*Note*: Page numbers followed by "f" indicate figures.'

A

Actinobacteria, 133–134
Active sites, 46, 67–68
Amino acids, 7, 25
Ancient evolution, 13, 49
Ancient mesophilic archaea, 136–137
Ancient proteins, 19
Anti-evolution propaganda, 173
Antiparallel twins, 59
Anti-religion, 51–52
Archaea, 76, 109, 109f, 123–124, 161
Archaeal RNA polymerase, 123
Archaeal transcription, 115, 116f
Asgard archaea, 133
Atmosphere, 129, 155
Atmospheric CO_2, 149
ATP hydrolysis, 83

B

Bacteria, 109–110, 110f, 161
    bacterial RNA polymerase, 126
    bacterial σ factors, 125
    Escherichia coli σ 70, 125
Bacterial endosymbiont, 147
    engulfment, 135, 135f
Bacterial RNA polymerase, 65
Bacterial sigma (σ) factors, 29, 125
Bacterial transcription, 117, 117f
Biological complexity, 107
Biosphere, 2, 130, 155
BRE_{down}, 123
BRE_{up}, 123
Bridge helix (BH), 68

C

Caenorhabditis elegans, 158
Calvin cycle, 148
Cancer databases, 175
Carbon dioxide (CO_2), 148
Carboxy terminal domain (CTD), 79, 140, 174
    consensus repeats, 142
    extensive CTD interactome, 142

phosphorylation, 157
    RNA polymerase II, 141–142
cBioPortal cancer database, 158–159
Cell nucleus, 2
Central Atlantic magmatic province (CAMP), 151
Chaotic genetic mechanisms, 49
Chaparonin proteins, 47
Chemical polymers, 83
Chemical synthesis of life, 83–84
Chimera, 21
Chloroplast/plastid, 2, 147
Citric acid cycle, 83
Cloverleaf tRNA, 28–29, 28f, 87, 177
Coevolution, 95–96
Coevolutionary forces, 89
Compartmentalization, 85
Computers, 53–54, 175
Core protein folds, 49
Core protein motifs, 4, 13, 16f
    RNA ligation, 17
Cotranscriptional splicing, 155
Cradle-loop barrels, 59, 60f
    RIFT barrel, 60–61
    swapped-hairpin barrels, 59–60, 61f
Crenarchaeota, 145
Cretaceous-Paleogene Extinction, 153–154
Criticism, 9
Cryoelectron microscopy, 169–170
CTD. *See* Carboxy terminal domain (CTD)
CTD interactome, 157–160
Cutout doll problem I, 11–16
Cutout doll problem II, 17–20, 18f–19f
Cutout doll problem III, 59–64
Cyanobacteria, 129
Cyanobacterial DNA, 147
Cyclin domains, 155

D

DDRPs. *See* DNA-dependent RNA polymerases (DDRPs)
Deccan Trapsvolcanoes, 153
Denisovan DNA, 161
Dimeric RIFT barrel, 59–60

Dimeric swapped-hairpin barrels, 59
Divergence, 14, 76
DNA, 7, 25, 27, 55
DNA-dependent RNA polymerases
    (DDRPs), 29–31
DNA genomes, 3, 63, 75f
DNA integration, 27
DNA replication, 27
DNA-template-dependent multi-subunit RNA
    polymerases, 14, 63
DNA-template-dependent RNA polymerases,
    3, 63
DNA templates, 29, 30f
DNA topoisomerases, 113
Double-Ψ–β-barrels, 59, 67–68, 67f, 139
DPBB enzymes, 28
Drosophila melanogaster, 158

E
Endogenous retroviruses (ERVs), 162
Endosymbiosis, 1, 14–15, 136
Endosymbiotic fusion, 55, 57, 145
Energy generation, 85
Energy transduction, 1, 83
Enzymes, 29
Escherichia coli σ 70, 125
Eukaryotes, 14–15, 55, 110–111, 111f, 136,
    161, 174
    carboxy terminal domain (CTD), 81
    endosymbiosis, 15
    mold *Neurospora crassa*, 63
    RNA polymerases, 76–77, 77f, 141
    TBP-associating factors, 143
Eukaryotes complex, 2
Eukaryotic introns, 146
Eukaryotic multi-subunit RNA polymerases,
    139–144
Eukaryotic RNA polymerase II, 139–140, 140f
Euryarchaeota, 145
Evolutionary defense, 155
Evolution of life, 55

F
α/β Folds, 45–48
Francis Crick adaptor, 93–94

G
Gene fusions, 17
    rules of, 23
General transcription factors (GTFs), 108,
    115–122, 139–144

Genesis, 9
Genetic code, 25, 27
Genetic duplications, 17
Genetic errors, 17
Genetic repetitions, 17
Genomes, 14–15
Genomic data, 53
Geosphere, 2
Global positioning device, 155
Global warming, 149
Glyoxylate cycle, 83
Great Anthropocine Extinction, 171
Great Oxygenation Event (GOE), 129
Group II intron, 14–15, 31
GTFs. *See* General transcription factors
    (GTFs)
GTP hydrolysis, 83

H
Hatred of evolution, 51
α-Helices
    solubilize β-sheets, 45
    turns and coils, 46
Hepatitis- δ virus, 63
Homo ancestors, 161
Homology modeling, 169–170, 170f
Horizontal cellulase gene transfer, 153
Horizontal gene transfer, 2, 15
HTH (helix-turn-helix) factor, 29
Human biosphere, 65
Human cancer
    blood, 166
    cBioPortal website, 167
    DNA sequencing, 165
    DNA topoisomerase inhibitors, 166
    metastasis (cancer spreading), 166
    mutations, 165–166
    retinoblastoma (RB) protein, 166–167
    smoking tobacco, 166
Human evolution, 161–164
Human extinction, 171–172
Human RNA polymerase III, 169–170
Human transposable elements, 162, 162f
Huronian Glaciation, 129

I
Industrial Revolution, 171
Intelligent design, 23–24
Irreducible complexity, 161
Iterated pattern recognition, 51
Iteration of sequences, 111

K

Korarchaeota, 133–134

L

Last eukaryotic common ancestor (LECA), 31
    actinobacteria, 133–134
    ancient mesophilic archaea, 136–137
    archaeal genome invasion, bacterial group II
        introns, 136, 136f
    bacterial endosymbiont engulfment,
        135, 135f
    eukaryotic common ancestor, 133
    group II intron elements, 137
    late stage LECA evolution, 133–134, 134f
    Lokiarchaeota/α -proteobacteria, 133–136
    Thaumarchaeota and Korarchaeota, 133–134
Last universal cellular ancestor (LUCA), 1, 73,
        107–108, 108f
    ancient evolution, 14
    divergence, 14
    DNA genome, 13–14, 56
    DNA ligation and recombination
        mechanisms, 114
    DNA replication, 27, 56, 118
    DNA templates, 29, 30f
    DNA topoisomerases, 113
    general transcription factors (GTFs), 108
    promoter DNA sequence, 29
    reverse transcriptase (RTase), 108
    RNA-protein, 14
    transcription factor B (TFB), 108
    universal cellular common ancestor, 133
LECA. *See* Last eukaryotic common
        ancestor (LECA)
LEGO (Trademark) life, 19, 49, 63, 123
    DPBBs, 28, 28f
    transfer RNA (tRNA), 27
Lokiarchaeota, 31, 56–57, 110–111, 134–135,
        145–146
Long interspersed nuclear elements
        (LINEs), 162
LUCA. *See* Last universal cellular ancestor
        (LUCA)
LUCA general transcription factors/promoters,
        115, 116f
LUCA promoter, 115, 120

M

Man-made global warming, 172
Matryoshka, 147
Membrane, 85

Mesophilic archaea, 124
Messenger RNAs (mRNAs), 79
Metastasis (cancer spreading), 166
Methane, 129–130
Methanogenic archaea, 153
Molecular biology, 21, 22f
Molecular coding, 83
Molecular graphics, 7, 9–10, 21, 53, 178
Monomeric double-Ψ–β-barrels (DPBB),
        59–61, 60f
Monomeric RIFT barrel, 59–60, 60f
Multimerization, 17
Multiple endosymbioses, 133
Multiple extant archaea, 145
Multi-subunit 2-DPBB type RNA polymerases,
        77–79, 78f
Multi-subunit RNA polymerases, 14, 61–63,
        62f, 64f, 71, 73–82
    bacterial RNA polymerase, 65
    bridge helix (BH), 68
    double-Ψ–β-barrels, 67–68, 67f
    Thermus thermophilus multi-subunit RNA
        polymerase, 65–66, 66f
    trigger loop (TL), 68
    Zn1 and Zn2, 69–71, 70f
    σ factor binding to promoter DNA,
        68–69, 69f
Multi-subunit two double-Ψ–β-barrel-type
        RNA polymerases, 81

N

NCBI online resources, 53
Neanderthals, 161
Negative supercoiling, 120
Nucleic acids, 25

O

Ocean acidification, 149, 172
Online cBioPortal Cancer Genomics
        Database, 53
Oxidation–reduction power, 83
Oxidoreductases, 29
Oxygen, 129–130
Oxygen catastrophe, 129

P

Paleocene-Eocene Thermal Maximum,
        155–156
Pangaea, 151
Patchwork eukaryotic phylogenomics,
        145–146

PDB files, 21, 53, 71
Permian-Triassic Extinction, 149–150
Phagocytosis, 136
Phosphorylation, 157
Photosynthesis, 130
pH unit, 171–172
Phylogenomics, 3
Phyre2, 169
Plants, 147–148
Plants fix CO_2, 148
Plasmids, 147
Polymers, 25–26
Preindustrial CO_2, 149
Preindustrial times, 155
Pribnow box, 110
Primordial promoter, 119
Primordial TBP, 119
Prokaryotes, 161
Promoter-proximal pausing mechanism, 157–160
Promoters, 29, 115, 119
Protein catalysts, 3
Protein chemistry, 9
Protein cofactors, 89
Protein Data Bank (PDB) codes, 178
Protein design, 23
Protein pockets, 46
Proteins, 25, 74–75
α/β Proteins
    ABC ATPase protein, 40–41, 41f
    ancient (β– α)₆ domain, Swi-Snf ATPase
        RapA, 41, 42f
    ancient (β–α)ₙ repeat proteins, 43, 43f
    Aquifex aeolicus, 39, 40f
    Gallus gallus (chicken) triose phosphate
        isomerase, 35–36, 35f
    Homo sapiens estrogenic
        17-betahydroxysteroid
        dehydrogenase, 37–38, 37f
    PDB 8TIM, 36
    RCSB PDB website, 36
    rearranged Rossmann-derived fold protein,
        41, 42f
    second ancient (β– α)₆ domain, Swi-Snf
        ATPase RapA, 41, 42f
    Synechocystis Sp. glutamate synthase,
        35–36, 35f
    TIM barrels and Rossmann folds, 33, 34f
    TOPRIM domains, 33–34
    triose phosphate isomerase, 36–37
    β-keto-acyl carrier protein reductase, 38, 39f
    β-sheets, 33–34
Protein sequences, 7
Protein structure/function, 13, 46

Protein visualization program, 21
α-Proteobacterium, 14–15
Proton gradient, 85
Proto-ribosomes, 1
Pseudoknot, 60–61
Pseudosymmetry, 107
Pymol, 21, 71

R

R2 Cancer Database, 53, 165
RCSB protein data bank, 53
Repeating sequences, 25–26
(β–α)ₙ repeat proteins, 46
Replication, 29–31
Replication errors, 90
Replication origins (ORIs), 29–31
Retinoblastoma (RB) protein, 166–167
Retroviral replication, 27
Reverse transcriptase (RTase), 27, 108
Reverse transcription, 27
Ribosomal RNA (rRNA), 27
Ribosomes, 1, 7, 27, 74–75, 74f, 91
    peptidyl transferase center, 88
    primitive ribosome/proto-ribosome, 85
    RNA template, 87
    transfer RNAs (tRNAs), 87
Ribozymes, 14, 25, 74–75, 85, 89
RIFT barrels, 28, 59–60, 89
    cradle-loop barrel, 60–61
    dimeric RIFT barrels, 61
    monomeric RIFT barrels, 60–61
RNA, 7, 90
    genetic code, 25
    hydrolysis, 75–76
RNA catalysts, 74–75
RNA elements, 2
RNA-encoded protein synthesis, 3–4
RNA enzymes, 25
RNA genomes, 63, 75f, 90
RNA ligation, 17
RNA polymerase II, 57, 77–79, 79f,
    141–142, 141f
RNA polymerase III, 77–79
RNA polymerases, 3, 73
    DNA template, 67
    eukaryotes, 76–77, 77f
    RNA polymerase active site, 67, 67f
    two double-Ψ–β-barrel types, 62
RNA-protein world, 83, 89–90
RNA synthesis, 3, 7
RNA-template-dependent RNA polymerases,
    3, 14, 62

RNA-template-dependent two double-Ψ–β-
    barrel-type RNA polymerases, 59
RNA templates, 30f
RNA world, 85–86
Rossmann folds, 29, 29f, 89
    Homo sapiens estrogenic 17-betahydroxys-
        teroid dehydrogenase, 30f, 31
    β-keto-acyl carrier protein reductase, 30f, 31
rRNA. *See* Ribosomal RNA (rRNA)
RuBisCo, 148

S
Sandwich barrel hybrid motif (SBHM), 76–77
SBHM. *See* Sandwich barrel hybrid motif
    (SBHM)
Sea-level rise, 155
Selfish DNA, 2
Sequence repeats, 103–104
Sequestration, 85
β-Sheets
    aggregation of proteins, 45
    amino acid compatibility, 47
    C-terminal end, 46
    N-terminal end, 46
    turns and coils, 46
Small interspersed nuclear elements
    (SINEs), 162
Snowball earth, 129
Stamp collecting, 51, 175
Swapped-hairpin barrels, 59–60, 61f

T
TATA box, 110, 120, 123
Tectonic violence, 151
Templated protein synthesis, 4–5
Thaumarchaeota, 133–134
Therapsid animals, 150
Thermus thermophilus multi-subunit RNA
    polymerase, 65–66, 66f
TIM barrels, 29, 29f, 89
    Gallus gallus (chicken) triose phosphate
        isomerase, 28f, 29
    Synechocystis Sp. glutamate synthase, 29,
        29f

TOPRIM domains, 33–34
    Aquifex aeolicus, 40f
    β-sheets and α-helices, 39–40, 40f
Transcription, 29–31, 91
Transcription factor B (TFB), 108
Transfer RNAs (tRNAs), 4, 27, 85, 87,
        104–105
    2-amino acylated tRNAs, 28–29
    adapter, 93
    anticodon loop and T-loop, 91, 96f–97f,
        98–99
    bound to mRNA, ribosome, 98–99, 99f
    cloverleaf tRNA, 28–29, 28f, 92, 92f, 103
    D-loop, 91, 97
    free tRNA, 98–99, 99f
    intellectual property, 104
    nucleotide proto-tRNA microhelix, 94, 94f
    proto-tRNA minihelix world, 101
    proto-tRNAs and acceptor stems, 93–94, 93f
    ribosome, 102
    RNA coding of proteins, 28–29, 28f
    sequence logos, 95–96, 95f
    sequence repeats, 103–104
    V-loop, 95–96
Translation, 27, 91. *See also* Transfer RNAs
        (tRNAs)
Triassic-Jurassic Extinction, 151–152
Trigger loop (TL), 68
Triose phosphate isomerase, 36–37
tRNA. *See* Transfer RNA (tRNA)
Two double-Ψ–β-barrel-type RNA
        polymerases, 63, 64f
Type IIA DNA topoisomerase, 39–40

V
Viruses, 169
Visual Molecular Dynamics (VMD), 21

W
Winged helix turn helix (WHTH) proteins, 123
Wood–Ljungdahl pathway, 83

Y
YASARA, 21

Printed in the United States
By Bookmasters